STATISTICAL THINKING

STATISTICAL

A STRUCTURAL APPROACH

THINKING

JOHN L. PHILLIPS, Jr.

Boise State College

W. H. FREEMAN AND COMPANY
San Francisco

A SERIES OF BOOKS IN PSYCHOLOGY

EDITORS

Richard C. Atkinson
Jonathan Freedman
Gardner Lindzey
Richard F. Thompson

International Standard Book Number: 0–7167–0832–9 (cloth)
0–7167–0831–0 (paper)

9

Library of Congress Cataloging in Publication Data

Phillips, John L.   1923–
Statistical thinking: a structural approach.
  Includes bibliographical references.
  1. Statistics.   I. Title.
HA29.P517   519.5   73-3035
ISBN 0-7167-0832-9
ISBN 0-7167-0831-0 (pbk.)

To Elaine

CONTENTS

PREFACE

The Purpose of This Book

This book is *not* intended to be the main text of a statistics course. In such a course it should be assigned *before* work on the main text begins. Used in that way, it could be immensely helpful in breaking the psychological barrier that stands between many students and statistics—especially if statistics is a required course for those students. The most frequent function of the book, however, will probably be as basic reading in the statistics or measurement sections of courses that have broader content domains—courses in the behavioral sciences or in professional curricula (business, education, social work, and so forth) that are related to the social sciences. It can also be used in any course that is concerned primarily with measurement (for example, Educational Tests and Measurements) and in which the instructor does not assume a prior acquaintance with statistics. Varying numbers of chapters may be used to fulfill those various functions, provided only that the chapters assigned are consecutive and begin with Chapter One, Introduction.

It is a book for beginners. Beginners, it seems to me, have special problems. I have elsewhere (*The Origins of Intellect*, W. H. Freeman and Company, 1969, p. 130) criticized Piaget for what I called his "recapitulation theory" of school learning. I suggested there that it would be extremely inefficient to require a high school physics student to "begin his study of mechanics, for example, at the sensorimotor level, continue through the preoperational and concrete operational levels, and only then be ready to deal with the material on a formal level."

Such a progression might be an inefficient way to teach physics, as the quotation implies, but only if students have already achieved a substantial mastery of physical concepts at each of the lower levels. In a new content area, it is often necessary to drop at least one level below that of which the student is presumed to be capable in order to establish a solid foundation on which to build. That is, I believe, a principle recognized by experienced teachers of all theoretical persuasions.

In the teaching of statistics to college students who have virtually no background in the discipline, it is appropriate to begin with concrete images. The early chapters of this book, therefore, are liberally laced with such images. Only gradually is the student brought to the second-order abstractions of Chapter Eight.

Even then, he is invited to retreat occasionally—to consolidate relevant learnings from earlier topics and to build his confidence for the ones ahead. Many otherwise competent college students are frightened into confusion when confronted by the necessity of thinking quantitatively; they have had unsuccessful experiences with such demands in the past. But much of their travail has been related directly to the manipulative aspects of quantitative thinking—aspects that have been emphasized in their experience at the expense of logical structures.

The possibility exists that *such students might comprehend quantitative concepts if manipulations were deemphasized and if the logical development of concepts were carefully done* in the manner outlined above.

Of course, if one is training *practitioners* in statistics and measurement, manipulations are important in themselves; but students who will be statistical "consumers," so to speak, need to *understand* basic concepts, that they might know what the manipulations are *about* and that they might comprehend the results of manipulations done by others. And of course even the manipulatory skill of the practitioner must be built upon a solid understanding of the structure that underlies every manipulation.

This book can therefore function effectively in two ways. First, it can present basic ideas in the form in which they are needed by "consumers" of statistics. Secondly, for the relatively few who are destined for more advanced training in statistics, it can provide a helpful framework for the more detailed study that lies ahead.

The metaphor "framework" is apt, I think; for the approach of this book is "structural." It is concerned with the binding together of ideas —with the relation of each idea to others that the student has brought with him or that have been presented earlier in the text. There are frequent cross references, and details are subordinated to overarching principles. Indeed, it was the lack of structure in available texts that impelled me to begin working on this one in the first place. My colleagues and I were looking for a small book to accompany a two-week "measurement" unit in an educational psychology course. We found several paperbacks on statistics; but the only ones that were sufficiently basic for our purposes were programmed, and programmed texts have difficulty with structure—a difficulty that has been well stated by S. L. Pressey as part of his argument that such devices should be used as adjuncts to, rather than replacements of, more traditional methods of presenting ideas:

> The student is shown this material one bit or frame at a time in the window of a mechanism or space of a programmed textbook. He cannot readily look back at what he has been over or ahead to sense what is to come, or discover any outline or structure in the material. . . . For effective reading, for general understanding of main ideas, and for adequate study and review, this procedure seems to be as clumsy as asking a person to apprehend a picture by letting him see, in a set order, only one square inch at a time. . . .
>
> Study of a complex and structured subject seems better begun by an overview of reading matter to display the structure and order the complexity. A good book will show its structure in the table of contents and catalog its contents in the index; with such aids the learner can easily move about in its numbered pages with only the flick of a finger, using page headings and subheads in the text to guide him. He may turn back and forth from table or graph to related text, skip something already known, review selectively for major and difficult points. . . . ["Basic Unresolved Teaching-Machine Problems," *Theory into Practice*, 1962, vol. 1, pp. 31–33]

It has been my intention to write a "good book" in precisely that sense.

Some Specific Suggestions

The average college student, if he is motivated to succeed and confident that he will, should comprehend most of this text entirely on his own. There are, however, a few key ideas that are exceptionally difficult and concerning which even a good student might appreciate some help.

Confidence levels

The section that introduces the concept of confidence levels (pages 67–70) is difficult because the student has to (1) imagine a true distribution somewhere other than the location of the obtained distribution and then (2) find the probability of obtaining what he has indeed obtained. That is a reversal of ordinary patterns of thought in which established fact constitutes the standard against which other possibilities are judged. Because the confidence-level operation is such a strange mental maneuver, I have provided in the body of the text (pages 69–70) some suggestions for practicing it. The instructor can help by assigning the exercises and requesting answers (that is, probabilities acquired by inspection) and then displaying the same operations via overhead projector, using a translucent "distribution cutout" so that the vertical line above the obtained mean remains visible as the distribution is moved about on the base line.

The z ratio

The section on the z ratio is difficult because there are so many steps in its logical development—each one further than the last from the concrete empirical fact of test performance. Here the teacher can help by tracing the steps carefully, beginning with the selection of two sets of sample means from a population (of French grammar students all of whom have had the same quality of instruction), and continuing through all the diagrams in Figures 8-1 through 8-3 to the placement of the obtained difference within a distribution of differences in Figure 8-4. The first (population) diagram might be drawn on a chalkboard to emphasize its size relative to the others, and the latter might be reproduced on a transparency for overhead projection. In any case, what the teacher can add to the text is *movement*—an emphasis on the transformations that occur from one diagram to the next.

Analysis of variance

In many courses, assignment of the section on analysis of variance will not be appropriate. Those students who do progress that far will find that the central problem is the extraction of *variance among means* from *total* variance. It might be helpful to demonstrate the component character of each individual score by analyzing several such scores in the context of Figure 9-1. The teacher may choose any point on the baseline as a score to be analyzed, designate that score as a part of one of the six distributions, and then show how the deviation of that score from the grand mean can be described as the algebraic sum of (1) its deviation from the mean of its own distribution and (2) the deviation of that mean from the grand mean. The process may be repeated for any value on the scale.

While Figure 9-1 is still on display, the teacher might show with a gesture how the elimination of variability among means would result in a collapsing of all six distributions into the single distribution portrayed in Figure 9-2. As a preview of the section on analysis of variance, that (plus the ensuing discussion) might save students both time and frustration. Again, what a text cannot do is to *move*, and transitions are best communicated by movement.

Acknowledgements

Many people have helped in various ways with this book: my colleagues Virginia Cox and Jason Smith (anthropology and archeology), Ellis Lamborn and Steve Hamilton (business and economics), Pat Bieter and John Jensen (education), Angus McDonald (education and psychology), Willard Overgaard and William Skillern (political science), Wylla Barsness and Delbert Heacock (psychology), James Christensen, Robert Corbin, Patricia Dorman, and Carol Harvey (sociology), Irene Wilcox (social work), and the many students who criticized its prototype.

I want especially to acknowledge the assistance of Mark Snow and Steve Thurber of the Psychology Department for their critical readings and for the many discussions we have had concerning the technical aspects of the book.

JLP

Boise State College
November 1972

STATISTICAL THINKING

ONE

INTRODUCTION

You are probably taking up statistics not because you want to but because you have to. If so, I know how you feel. I went through the same experience years ago; if I could have avoided statistics, I probably would have. However, my attitude changed after I began to study it, for I discovered in it a new way of thinking that really fascinated me.

But your task is in an important way more difficult than mine was. You won't have to do the computations that I did, but you are being asked to acquire in a fraction of one semester (or quarter) an understanding of the same basic concepts to which I devoted two full semesters.

The Task

Your plight and your prospects are well illustrated, I think, by the experience of a student who was asked to evaluate the prototype of this book:

It was the most difficult book I have ever read. It was foreign to me since I had had no background knowledge of the things talked about . . . If I was to understand it, I realized I would need to outline the book chapter by chapter. I did so, and to my amazement, it followed a very orderly pattern after all. It really did present what the author had stated he hoped to present. If you understood Chapter One, you could see the logic of Chapter Two, and so on through each chapter. I feel I learned a great deal about measures in a relatively short period of time.

That quotation also contains some important advice on how to use the book. There's only one thing I'd like to add to it in that regard: even though you may have mastered the ideas that precede the one you are working on at any given moment, you should be prepared to go back to those ideas from time to time and consider their relation to the new one that is being presented. I have tried, by providing frequent cross references, to help you do just that. (You may wish to keep a couple of bookmarks handy for this purpose.) When you are finished, you should have in mind a hierarchical *structure* such that each new idea is related to one or more of those that have preceded it.

The acquisition of such a structure is highly satisfying in itself, but that probably is not the reason that a unit on statistics has been included in the course in which you are now enrolled. Why, then, has it been included? An understanding of statistical concepts will not be of critical assistance to you in gathering economic data, in conducting interviews for a political or sociological opinion or attitude study, in uncovering archeological artifacts, or in teaching children. But often people who do economic, political, sociological, anthropological, archeological or educational studies (to name but a few) report their findings in statistical terms. If you are planning a career in any of the several professions to which those studies are relevant, it is important that you be able to read them with understanding. A continuing acquaintance with developments in his field is the mark of the true professional, and this book will help you to maintain that acquaintance.

The Basic Ideas

If you want to understand the meaning of any measurement in the social sciences, you must come to know at least two things about it. First, you must be able to describe the operations by which it was obtained, and second, you must be able to place it in relation to other measurements that have been obtained in the same way. (You may also find it helpful to place it within some theoretical framework, but not everyone agrees on that.)

This book is concerned primarily with the second kind of knowledge. Statistical thinking deals with multiple measurements. It analyzes the relation of "how many?" to "how much?"—of frequencies to scores. If the basic element in measurement itself is a *score*, the corresponding concept in statistics is a *distribution* that includes many scores—in short, a *frequency distribution*.

Such a distribution can be described by drawing a picture of it—and indeed in the early stages of your learning about distributions, that is the method I shall use to describe them. The method is cumbersome, however; so ways have been devised to achieve roughly the same result through the use of numbers rather than diagrams. The most important advantage of numbers over diagrams is that they can be manipulated in a way that diagrams cannot.

Imagine that (for reasons known only to your psychoanalyst) you have just had a pile of gravel dumped on your front lawn and that (for similar reasons) you want to describe the result to me over the telephone. To do that successfully, you will have to tell me at least three things about the pile: (1) its general configuration—that is, whether it is shaped like a cone, a pancake, or your garage roof—(2) its location—that is, how far and in what direction it is from some reference point that is familiar to me—and (3) the extent to which it is "spread out"—for example, if it is a cone, is it a steep-sided one that covers only a small area, or is it a low-profile cone that covers most of the lawn?

That pile of pebbles is analogous to a frequency distribution of scores, and the same kinds of information are needed to describe either adequately. Concerning configuration, we have certain names for frequency distributions—names that convey information to anyone who is familiar with them—names such as "normal," "symmetrical," "positively skewed," and "bimodal." Concerning location, the procedure is pretty much the same as it is for a pile of gravel. A dimension is identified and a reference point is chosen; measurement of distance to the center of the distribution is from that reference point. The result is a *measure of central tendency*. And finally, to communicate information about the amount of dispersion, a new reference point is used—namely, the center of the distribution—and the needed information may consist of the average distance of individuals (like that of the individual pebbles in a pile of gravel) from the central point. That distance is a *measure of variability*.

But just describing a distribution is not always enough. Often you will be interested in *two* distributions and in the relationship that exists between them. Consider a single variable, "IQ in the general population," and a few other variables with which it might be paired: IQ and family income, IQ and some index of health care, IQ and a general index of socioeconomic status, IQ and race, and IQ and place of residence. Other interesting relationships could be investigated among the IQs of various subgroups of the general population: between parents and their children, between identical twins, between fraternal twins, between non-twin siblings, and between pairs of unrelated children. Those are just a few that come to mind at the moment. Others would

occur to you if you were making a study of intelligence and its correlates, and in every case you would need a way of communicating your
findings to others; in short, you would need a *measure of relationship*.

Whether or not you wish to relate one set of scores to another, you
will surely want to be able to tell what each individual measure means.
If you are a teacher who has given my child a test, and I ask you how
well he did on it, you might try to put me off with a "high" or a "low"
and go on to talk about something else. But if I want to know *"How
high?"* you are in trouble. You may answer that he got 90 percent of
the test items right. You think you are off the hook, but I persist:
"How *hard* is that test? 'Ninety percent' is very impressive if the
items are all difficult, but not if most of the other kids scored above
ninety-five!"

Our conversation has been concerned entirely with the *interpretation
of individual measures*, and I'm sure you'll agree that all of my questions are pertinent. Without answers to them and others like them,
one really cannot know the meaning of a score.

On the other hand, you must be careful to avoid *over*interpreting
measures, whether they be of individuals or of groups. If, for example,
you were to weigh a *random sample*[1,2] of 50 ten-year-old boys, could
you easily compute a measure of central tendency for the *sample* that
is identical to the central tendency of a population that includes *all*
ten-year-old boys? How different might the obtained average be if it
were computed from a *different* random sample of that same population?
Questions like these have to do with *precision of measurement*, and we
do have ways of dealing with them.

---

[1] A random sample is one in which (*a*) every member of the population has an equal
chance of being included in the sample and (*b*) each selection is made independently
of all the others.

[2] The notes in this book are of two kinds and are denoted by two kinds of superscript.
Footnotes are indicated by numbers ([1], [2], [3], etc.); other notes, which appear in the
back of the book, are indicated by asterisks (*).

The footnotes are intended to be supplementary to the central discussion at any
given time and are important to a full understanding of that discussion. A footnote
may provide (*a*) refinement of an idea by restricting or extending its implication, (*b*)
explanation of a pedagogical technique by commenting on the relative importance of
its various attributes or (*c*) cross references that are not essential but are enriching.
In fact, the objective in every case is enrichment. The way to use the footnotes, then,
is (1) *to ignore them* as you work through a section or chapter for the first time and
(2) *to study them carefully* the second time through. Every chapter has a few footnotes.

The notes in the back of the book are more technical and go beyond the scope of
the text. Ignore them at least until you have mastered the chapter to which they refer.
These notes pertain to subjects mentioned in Chapters Three, Four, Five, Eight,
and Nine.

Assuming for the moment that you do know how to deal with questions about precision, consider this one additional question: Upon further analysis of your sample of ten-year-old boys, you discover that there is a marked difference between the weights of boys living in one area of the country and those living in another. You suspect that the difference is due to diet, and you subsequently narrow that hypothesis down to a single vitamin that seems to be more plentiful in one of the areas than in the other. One way to test your hypothesis would be to select two samples of male infants who are living in the area in which the vitamin is least plentiful, introduce the suspected vitamin into the diet of one group, and then after a period of nine years weigh them both again. Let's imagine that there is a difference between the two. Is it large enough that you can be reasonably sure that it did not occur by chance—that a replication of the same study would not turn up a difference of zero, or even a difference in the other direction? To put it another way, "How *significant* is the difference you obtained?" Again, there are ways of dealing with such questions.

In the chapters that follow, each of the above ideas will be developed further, but always in the same vein as those you have seen here. If you are a sophisticated mathematician—or a frustrated accountant—this book is not for you; it is aimed directly at the underlying *logic* of statistical thinking, with an absolute minimum of arithmetical and algebraic manipulations. You will find the logic similar in many ways to common sense. The main difference is that it is rigorously systematic, and like any system, its parts are interdependent. So the book cannot be studied piecemeal; the ordering of the chapters was deliberate and necessary. Once you have worked through it, however, the book will serve as a convenient reference to take with you into your professional life. It has been organized with that in mind.

FREQUENCY DISTRIBUTIONS

Every year, Cerebral University administers a "scholastic aptitude" test to each of its incoming freshmen. After the tests have all been scored, each freshman is handed a small card with his score on it; then all the freshmen are herded into the seating area of the football stadium.

A university official stands on the sideline of the football field with a microphone in his hand. He points to the space between the west goal and the adjacent five-yard line and announces that anyone with a score below 5 should come down and stand in the middle of that space. Nobody moves, so he walks five yards to the east and repeats the instructions for that space. There is a long pause, then one miserable soul slinks down to the field; he is the only one in the *class interval* 5 through 9.[1] Another call (10 through 14) yields two students, a fourth

---

[1]Technically, "5 through 9" is only an *apparent* class interval. Its precise limits—often called "true limits"—are 4.5 and 9.5. Those of the next higher interval are 9.5 and 14.5, and so on up the scale. This adjustment of the class intervals is required because test scores are given by discrete or unique integers such as 5, 6, 7, 8, and 9, while the precise limits refer to the continuous intervals extending from 4.5 to 5.5, 5.5 to 6.5, 6.5 to 7.5, 7.5 to 8.5, and 8.5 to 9.5. The continuous interval 4.5 to 9.5 is the sum of all those.

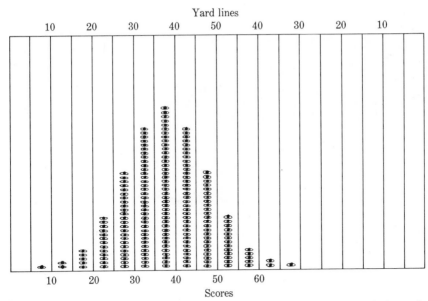

FIGURE 2-1. Students standing on football field in columns designated by aptitude test scores.

(15 through 19) gets him five, a fifth (20 through 24) 13, and so on, with increasing *frequencies* up through the middle scores, followed by decreasing ones after that. Figure 2-1 is an aerial view of the field after all of the students have left their seats.

Normal Distributions

Cerebral U. lacks the budget for it, but if our official had a rope long enough to run from the sideline, out around the entire freshman class, and back to the sideline (and if the official were indeed to use it in that way) the rope would form what is known as a *normal curve.*[2] It would be so called because it would enclose a *normal distribution.*

Many quantities that are complexly determined do tend to form normal distributions. Scholastic aptitude measures often show that tendency. Such measures have many determiners, both hereditary and

---

[2]It is sometimes called a *bell* curve, because of its bell-like shape. If instead of draping the rope loosely in a smooth curve, our official were to instruct the end student in each column to grasp it firmly and pull it taut, the rope would form a series of straight lines and angles known as a *frequency polygon.*

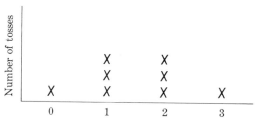

Number of heads in a single toss of three coins

FIGURE 2-2. Distribution of eight coin tosses.

environmental; some of those determiners influence an individual's score in an upward, some in a downward direction. In a few individuals, there is a preponderance of positive determiners; theirs are the scores in the upper (right-hand) *tail* of the distribution. In a few others, negative determiners predominate; their scores form the lower tail. Most scores, however, represent more balanced combinations of determiners; they form the large middle area of the distribution.

All of this is assumed to be a matter of chance, and you can readily see that the probability of, say, 100 determiners being all positive (or all negative) in any individual would be vastly smaller than the probability of their being approximately half and half. If you don't believe that, toss a mere three coins[3] just eight times[4] and plot the number of occurences against the numbers of heads that could possibly occur in a single toss (0, 1, 2, and 3 are all possibilities). You should come up with a distribution similar to the one in Figure 2-2. "All heads" (score = 3) does not occur often by chance, and neither does "all tails" (score = 0). The middle scores are much more frequent.

By the way, do you get the impression when looking at Figure 2-2 that if there were more class intervals and a larger number of observations the distribution of coin tosses would look very much like the approximately normal distribution of students that we got from Cerebral U.? If you do, you are right; the larger that number is, the more closely the actual distribution approximates the mathematical model we call "the normal curve." (This is the first of several references to effects of sample size. The symbol for the number in a sample is N; a *sample* is made up of the individuals or observations—that is, the number of *scores* under discussion. In the case of the CU freshmen,

---

[3] Three coins correspond to three determiners instead of the 100 cited in the example we just considered.

[4] Eight tosses correspond to eight students as compared to the 200 CU freshmen.

N is 200; in that of the coin tosses, it is 8. The relation of the sample to its *population* is discussed on pages 4 and 15–16.)

There are several ways in which a frequency distribution can deviate from normality, but we shall mention only four of them here: "positive skew," "negative skew," "J," and "bimodal."

Skewed Distributions

If the lower tail of a distribution is extended abnormally, that distribution is said to be *skewed*, and since the direction of the extension is downward it is further categorized as *negatively* skewed. If your instructor in this course should give you a very easy test, the distribution of your scores might look rather like Figure 2-3, which illustrates a negative skew.

If the upper tail is similarly extended, the skew is said to be positive, and the distribution is *positively* skewed. A difficult test would produce a distribution more like Figure 2-4, because only a talented (or industrious) few deviate very far from the lowest possible score.

FIGURE 2-3.   *Negatively skewed distribution.*

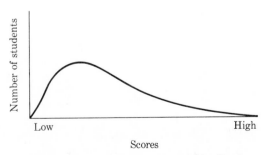

FIGURE 2-4.   *Positively skewed distribution.*

Other Configurations

There are other possibilities. If we were measuring conformity behavior, like that of motorists at a busy intersection, we might get a
J *curve*, like the one in Figure 2-5. The same would be true of a distribution of scores on a test that is extremely easy (so easy that most
people get perfect scores) or extremely difficult (so difficult that most
of the scores are zero).

An entirely different configuration would emerge if we were to
measure the standing height of humans; for there are two physical

FIGURE 2-5. J *curve of conforming behavior.*

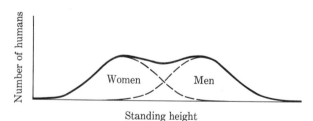

*FIGURE 2-6. Bimodal distribution of hu-
mans on a scale of height.*

types of humans—male and female—and that should result in a bi-modal distribution, as shown in Figure 2-6. Or, if your instructor were to spring a pop quiz on Chapter Five of this book at a time when only half of the class had read it, the distribution of those scores would be bimodal. (Note that in Figure 2-6 the distribution is apparently quite symmetrical; but a bimodal distribution can be asymmetrical, as indeed this one surely would be if, say, it were composed of twice as many women as men.)

But the "J" distribution is difficult to deal with statistically, and a bimodal distribution can be dealt with by separating the two normal distributions that are partially concealed within it. So these "other configurations" are of only passing interest to us here.

Summary

Many frequency distributions in the social sciences are approximately "normal"; that is, in the typical distribution there are a few very low scores and few very high ones, but the great mass of individuals tends to pile up on the middle scores. This occurs because the probability of all the many determiners of a trait pointing in the same direction is virtually nil, and that of a balanced combination of determiners is much higher. Although other configurations do occur (positive skew, negative skew, J curve, and bimodality are mentioned), our primary concern in this book will be the normal distribution, since it serves as a good approximation to the kinds of distribution frequently encountered in behavioral investigations.

MEASURES OF CENTRAL TENDENCY

Somewhere in the intermediate grades you were introduced to the concept of "average," and you have used it ever since on the assumption that there is only one such. Actually, there are several averages, of which three will be described here.

An average can represent either the sample itself or the *population* from which the sample is taken. A population comprises all the individuals who share certain specified characteristics. We might speak of the average scholastic aptitude score made by entering freshmen at Cerebral University, for example. If we actually have scores from only those freshmen who entered the University in a particular year—say 1970—our statement of the average for "entering freshmen" is based upon the assumption that the 1970 average is the same as the average for all freshmen who enter the University in all years, past and present.[1]

---

[1]Actually, of course, CU officials do not herd their freshmen onto the football field. They don't even use data from the current class. What they really do is to compare each new freshman's scores to those of last year's freshmen, on the assumption that there is no significant change in the entering freshman class from one year to the next.

On the other hand, we may use the average to represent only the class whose scores are on hand, in which case our sample average is precisely equal to the population average. Or we may use it to represent entering freshmen everywhere, in which case the sample average probably will *not* be equal to the population average. We proceed in each case as if our *obtained* average from the sample is also the *population* average. But that assumption is valid only to the extent that the sample really "represents" the population. In other words, the more we attempt to *generalize* from an obtained average (or any other statistic, for that matter), the less confident we can be of the accuracy of our statement.

But in science, generalization is the name of the game. There may well be some immediate practical reason for ascertaining the average scholastic aptitude score of freshmen entering Cerebral University in the current year (in that case, the sample would be identical to the population); but when a scientist observes a phenomenon in his laboratory, nothing of importance has been accomplished unless the result can be generalized. The phenomenon is a mere curiosity unless it can be replicated (for example, with subjects of a different species—or at least with other members of the same species). All science is based upon generalization; we almost never study all the individuals of a population in which we are interested. Instead, we study a sample, and one very useful description of a sample is its central tendency or average.

The Mean ($\overline{X}$)

The average that you learned about in grade school was a *mean*. The mean is not, as you were led to believe, *the* average; but it does have characteristics that make it the best one to use in many circumstances. Whenever the frequency distribution is fairly symmetrical and plenty of computation time is available, the mean is the statistic of choice. The computation time is necessary because all of the scores must be added together before the sum can be divided by N (number of scores), and there are sometimes thousands of scores to add. The formula for the mean is

$$\bar{X} = \frac{\Sigma X}{N} \tag{3-1}$$

where $\bar{X}$ is the mean, Σ is a combining term meaning *sum of*, and X refers to the *raw scores*[2] that have been recorded. The expression "ΣX"

---

[2] Raw scores are the only kind we have dealt with so far; others will be introduced later.

1 2 3 4 5 6 7 8 9 10

FIGURE 3-1. Fourteen cubes on a weightless beam.

(read "summation ex"), therefore, is the sum of all the raw scores in the distribution. Divide it by N and you have your mean.

I have given all this attention to the computation of the mean mainly because it gives me an opportunity to introduce you to a few of the symbols that we will need later. Actually, learning to compute a mean should receive very little of our attention for two reasons, the first of which is that you already know how to do it.

The second reason is that computation is not what this book is about; our objective is rather to acquire an understanding of statistical concepts. In the case of the mean, the following illustration should help to accomplish that.

Essence of the concept

Try to imagine a long beam made of an exotic metal that is (1) totally rigid and (2) utterly weightless. Upon that beam we shall place 14 cubes (to represent 14 scores), each of which has a weight equal to that of each of the others. Figure 3-1 shows one possible distribution of cubes.*[3]

At what point on the beam would a fulcrum (support) have to be placed to establish an equilibrium. That is, what is the *balance point* of the distribution?

This distribution is symmetrical, so it should be easy to see that it would balance if the fulcrum were placed at 5.5. But that is what all the computation is about; by using the formula, we can find the precise point we are seeking[4] even in cases in which a diagram would be rather

---

[3]Asterisks (*) refer to notes that appear at the back of the book and need not be read. They are there to provide an option for exceptionally motivated students but presumably will be omitted by most. Such a note may be more technical than is necessary for a clear view of the *structure* of a concept; it may introduce a term not used elsewhere or a related fact that is not explained in the text; or it may cite an unresolved controversy among statisticians. In any case, you may skip it with impunity.

[4]$\bar{X} = \dfrac{\Sigma X}{N} = \dfrac{77}{14} = 5.5.$

FIGURE 3-2. *Distribution of Nisei on an IQ scale.*

puzzling. That such situations do arise will become clear in our section on the median (pages 18–19).

In any event, a good verbal definition of the mean is "the *balance point* of the distribution."

Appropriate applications

The most important advantages of using the mean are that it (1) is the most stable measure of central tendency and (2) lends itself to the computation of other notably stable measures.

By "most stable" I mean that it varies least among repeated random samples selected from the same population. Let us say we are interested in finding the mean WAIS[5] IQ of second-generation Japanese-Americans (Nisei[6]). For simplicity, let us assume that the population itself remains stable during the time that it takes us to sample it. Such a population would probably be distributed normally; let's assume that it is and that Figure 3-2 is an accurate rendering of that distribution.

Theoretically, we could simply test every Nisei in the country, sum their scores, and divide by their N. Practically, however, that would be virtually impossible; as mentioned earlier, measurements are almost never taken of whole populations. What we actually do is to measure a *sample* of whatever population we wish to study. There are more ways than one of selecting a sample. Conceptually the simplest of them is *random* selection, in which the composition of the resulting sample is entirely a matter of chance. (See Footnote 1 on page 4.)

Now back to the stability of the mean. If we were to select, say, 1000 individuals at random from the population of Nisei, we could

---

[5]Wechsler Adult Intelligence Scale.
[6]Pronounced "Nee'-say."

FIGURE 3-3. *Superimposed distributions of population, sample, and sample means.*

μ_X = *Mean of the population (hypothetical).*
  (*A parameter.*)
$\bar{X}$ = *Mean of one sample (obtained).*
  (*A statistic.*)
$\mu_{\bar{X}}$ = *Mean of an infinite number of sample means (hypothetical)—the same as the mean of the population.*

compute a mean of that sample. If we should return that 1000 to the population and then select another sample of the same number in the same manner, we could compute a second mean. Returning those subjects to the population, we could select a third sample, and a fourth, and so on *ad infinitum*. Each sample would have a mean, and not all of those means would be the same. In fact, they would form a distribution—a distribution having the same bell shape as that of the population and of each of the samples.* It would, however, be a much more compact distribution than any distribution of individuals. Figure 3-3 superimposes distributions of the population, of a single sample, and of the means of many samples.

In Figure 3-3, the mean of the means ($\mu_{\bar{X}}$, or "Mu-sub-ex-bar") has been placed at the mean of the population (μ_X, or "Mu-sub-ex"), and indeed if an infinite number of means were taken, their mean would be the same as the population mean. But, of course, since that is never done, we can never be sure precisely where the population mean really is. (Remember, we know the population only through the samples that we draw from it.)

What we *can* do is to confine the population mean within certain limits; or more accurately, we can discover the limits within which the population mean probably lies. The horizontally compact configuration in Figure 3-3 is the distribution of sample means; when it is extremely narrow, *we know that the population mean is very near to our sample mean.*

At this point, let me clear up a possible misunderstanding. We do not actually take one sample after another and plot a distribution of their means; rather, we estimate mathematically[7] what would occur if we *were* to do that. Actually, we have only one sample—hence the reference in the preceding paragraph to "our sample mean"—and every bit of knowledge we acquire about the population comes necessarily from that sample.

We call any measurement that we make of the sample a *statistic*; the mean IQ of our sample of Nisei is one such statistic. The mean IQ of the population (all Nisei) is called a *parameter* of that population; any other characteristic of the population would be another parameter. The parameters are there whether or not we compute the statistics. In other words, every measurement (statistic) we obtain from the sample is an estimate of a particular population characteristic (parameter).[8]

All this is background essential to the understanding of the proposition that of all known measures of central tendency, the mean is the most stable—that it "varies least from sample to sample of a given population." If we were to take repeated samples of Nisei and compute any *other* single measure of central tendency for each sample, the distribution formed by those sample averages would be wider (less compact) than would a distribution of sample means. The mean varies less from sample to sample than does any other measure of central tendency. That is what is meant when we say that it is more stable.

The other significant characteristic of the mean is that it lends itself to the computation of other important statistics. Among those are the *standard deviation* (see Chapter Four, Measures of Variability), the *product-moment coefficient of correlation* (Chapter Five, Measures of Relationship), and all of the many *standard errors* (Chapter Seven, Precision of Measurement). None of them can be educed without first computing a mean.

---

[7]The information that is needed from the sample in order to make this estimate concerns its size and its *variability*. The variability of the sample is discussed in Chapter Four, Measures of Variability; while the estimate itself can be found in Chapter Seven, Precision of Measurement.

[8]See also the section titled "Standard Errors" in Chapter Seven.

The Median (Mdn)

Another way of indicating central tendency is to tell what point on the base line divides the distribution into two equal parts. Note that I said "divide the *distribution* in half, *not* divide the *base line* in half. Look back at Figure 3-1. There, the beam is the base line, and 35/2 = 17.5 is *not* the *median*. Nor is the median defined as a point halfway between the lowest point (1.5) and the highest (9.5), although because of the perfect symmetry of the distribution, it happens to be there in that particular case.

Essence of the concept

In *every* case, the median is the point between the lower and upper halves *of the distribution*. (Remember, the distribution is the group of individuals or scores comprising the sample.) In Figure 3-1, that point is 5.5, because there are 7 scores below and 7 above it. In *any* sample with an *N* of 14, the median will be midway between the 7th and 8th scores (counting up from the bottom of the distribution); in any distribution of 1000 individuals, the median will be the point midway between the 500th and 501st; and so on with as many illustrations as you care to cite.*

Appropriate applications

"Who is in what half of the distribution?" If that is our question, the answer will depend upon our first finding the median. A more important characteristic of the median, however, is that although it is not quite as stable as the mean, it can be used in situations where the mean would be inappropriate—namely, where the distribution is markedly skewed (page 9); where it is generally symmetrical but includes a few extreme scores at one end; where one end of the scale of scores is not long enough to reveal the true variability in the sample,[9] or where any combination of those conditions exists. It may not be too much of an oversimplification to say that the median should be used in preference to the mean whenever the shape of distribution departs radically from perfect *symmetry*. Consider the situation depicted in Figure 3-1. There, the distribution is nearly normal and perfectly symmetrical, so that the mean and the median are at precisely the

---

[9]An additional principle applies here. Because some of the scores are indeterminate, there is no way to know "how far out on the bar" the weight should be placed, and it is the placement of such weights that determines where the balance point is.

FIGURE 3-4. *Fourteen cubes on a weightless beam.*

same place (5.5). Now look at Figure 3-4 to see what happens when that symmetry is disturbed.

In Figure 3-4, we have exactly the same distribution as the one in Figure 3-1, except that two of the scores have been shifted far to the right. The balance point—that is, the mean—has shifted also; it is now 9.5, which, lying as it does above 12 of the 14 scores, is probably an inappropriate index of the central tendency of this distribution.

But what has happened to the *median* as a result of that shift of two scores? It hasn't moved at all! (Count the scores above and below it, and see for yourself.) You may say that it *should* have moved—at least a little bit—because the distribution is different than it was before; but, even though it is insensitive to that change, I'm sure you will agree that in this distribution the median is a better measure of central tendency than the mean, because the two extreme scores do have too much influence on the mean. Often, what has happened in such cases is that scores have been forced into one category when they should have been classified into two or more. For example, if you were to plot a distribution of the annual incomes of a football coaching staff, you would probably find that most of them are rather close together but that the salary of the head coach is distinctly separated from the others. If you had to report a single average of coaching salaries at Cerebral University, would you use the mean or the median?

And always remember that any statistic (in this case the median) is a *measurement* of a sample but only an *estimate* of the corresponding parameter.

The Mode

The last of the three averages to be presented here is also the easiest. The *mode* is simply the point at which the greatest frequency occurs.* In Figure 2-1, the mode is 37; in Figure 3-1, it is 5.5; in Figure 3-4, it is 5.0.

The mode's very simplicity explains one of its two main uses: it is used when a very *quick* estimate is needed. The other occasion for its use occurs when we are interested specifically in identifying the most common score.

With the exception of those two special situations, the mode probably should not be used at all, for it is more subject to chance fluctuations (it is less *stable*) than either the mean or the median, and it is the one measure of central tendency that may have more than one true value.

Summary

Statistics are numbers that carry information pertaining directly to *samples* and indirectly to *populations*. Unless an entire population is sampled, the sample may differ from the population; and since it is the latter in which we are interested, it is important to know which statistic is least likely to deviate significantly from the one we would obtain if we were to measure the entire population.

That statement is true of statistics of all kinds. If we are interested in "measures of central tendency," the *mean* is the statistic that meets our requirements—it is the one that varies least from sample to sample of any given population. The mean is also a companion to other stable statistics; in fact, it is a prerequisite to the computation of many of them. When the distribution is symmetrical and there is sufficient time for its calculation, the mean is the preferred measure of central tendency.

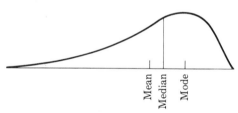

FIGURE 3-5. *Negatively skewed distri-bution.*

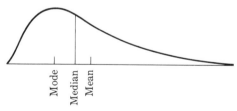

FIGURE 3-6. *Positively skewed distri-bution.*

The *median* is less stable than the mean (but more so than the mode). It is used in the few cases in which the information wanted is specifically at what point a distribution can be cut in half, but probably its major virtue is that it can be used with distributions that are not symmetrical. Unless the sample is very small, the median is also easier to compute than the mean—that is, the computation requires less time.

Finally, the *mode* is the quickest of all and the least stable. It should be used only as a preliminary estimate or as an index of "the most common score" or "the typical case."

In a skewed distribution, the arrangement (order) of the three measures along the base line is predictable. If the skew is negative, as in Figure 3-5, the arrangement is, from left to right: mean, median, mode. If it is positive, as in Figure 3-6, the order is just the opposite: mode, median, mean. Conversely, if you know the order of the three averages, you can tell the direction of the skew.

FOUR

MEASURES OF VARIABILITY

Different populations (and the samples extracted from them) have different central tendencies; but they differ in another significant respect. Consider the two curves depicted in Figure 4-1. Both represent distributions of the same area (identical N's), and both have the same central tendency; nevertheless, the two distributions are very different. In what way are they different? You can see that one is "spread out" more than the other. Since the base line on which the spreading occurs is a single scale of scores, the spreading means that the scores in that distribution *vary* more than do those in the "squeezed together" distribution. There are many occasions (some of which are discussed in Chapter Six, Interpreting Test Scores) when it is important to have some kind of numerical index of the variability of a set of scores.[1] In this chapter, we shall examine three such indices.

---

[1]Even if we had a diagram of every distribution drawn to scale so that we could compare variabilities by inspection, there would still be a need (and Chapter Six will demonstrate it) for an index that can enter into mathematical operations that are essentially numerical; you can't multiply or divide a visual perception with an acceptable degree of precision.

FIGURE 4-1. Two distributions with the same N's but different variabilities.

The Standard Deviation (S)

Taking its importance on faith for the moment, let us consider how an index of variability might be devised if we had none already available. It may help to have a concrete example in mind during the discussion, so let us imagine that the scholastic aptitude test mentioned in Chapter Two is actually a *pair* of tests—one of verbal, the other of numerical, aptitude. Their distributions might look something like Figure 4-2.[2] In that diagram, the means of the two distributions have been aligned so that we may concentrate on their respective variabilities. As we look at the figure, we are impressed immediately by the striking difference between the dispersions of the two distributions across their base line. But it is not enough to be impressed; we need a numerical index of dispersion (variability).

Essence of the concept

How might such an index be devised? One way would be to compare every score in the sample to every other and to average the differences obtained thereby. But that would be entirely too cumbersome, especially with large distributions; and we can get the same effect by

---

[2]The indicated central tendencies and variabilities of the two part-score distributions are such as to make somewhat plausible the proposition that when combined, they would form the distribution shown in Figure 2-1. However, the primary consideration in their selection was *simplicity*. We are deemphasizing computation; therefore, those computations that *are* required should be made as easy as possible. It is easier to think about an interval that extends from 20 to 25, for example, than one that extends from 23 to 27, or even from 23 to 28. You will find that you can do most— possibly all—of this book's computations entirely in your head.

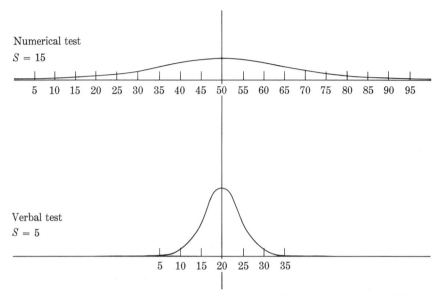

Numerical test
S = 15

Verbal test
S = 5

FIGURE 4-2. *Two distributions with the same N's but different variabilities.*

selecting a point of reference in the middle of the distribution and measuring the distance of each individual score from that point. The average (mean) of *those* differences (without regard to sign) could *also* serve as an index of variability. If, for example, we should choose the mean as our reference point, our index would be an average of the individual differences from the mean and could be obtained via the following formula:

$$AD = \frac{\Sigma|x|}{N} \qquad (4\text{-}1)$$

where AD = *average deviation* and $\Sigma|x|$ = sum of *individual deviations* from the mean of the sample.[3]

The "average deviation" *is* sometimes used as an index of variability, but it is not common enough to be discussed here strictly for its own sake. Rather, I have included it because it is a direct expression of the

---

[3]"x" (the difference between X and $\bar{X}$) is called a "deviation score." The vertical bars in "$|x|$" mean "without regard to sign." "Σ" is defined on page 13.

basic idea that underlies the most-used of all measures of variability. That idea is the concept of an *average of individual deviations* from the mean; that statistic is the *standard deviation.*

Now compare the little-used statistic that we have been discussing to the standard deviation (which comes very close to being *the* accepted measure of variability). Here again is the formula for the average deviation:

$$AD = \frac{\Sigma|x|}{N} \tag{4-2}$$

and here, the standard deviation:*

$$S = \sqrt{\frac{\Sigma x^2}{N}} \tag{4-3}$$

They are very similar, are they not? In fact, they are exactly alike except that for the standard deviation we take a mean of *squared* deviation scores; then we take the *square root* of that mean.[4] Again the important point is that the standard deviation is a kind of *average of individual deviations* from the mean of the distribution.

Appropriate applications

The standard deviation is to measures of variability what the mean is to measures of central tendency. In a normal distribution, (1) it is the most *stable* of variability indices, and (2) it lends itself to the computation of other notably stable statistics.

The sampling stability of the standard deviation can be conceived in the same way as that of the mean. (See pages 15–17.) That is, of all

---

[4]By now you may have suspected that any set of values of a given kind can be averaged. We have previously dealt with the mean of raw scores (pages 13ff) and of means (pages 15ff). Note that deviation scores, in a normal distribution, meet the requirement of *symmetry* that was set earlier (page 20) for the legitimate use of a mean. The vertical bars in the AD formula mean *without regard to sign;* a deviation score of "−10," for example, is at the same distance from the mean as a "+10." In the formula for the standard deviation, the squaring process has the same effect. The distribution of such deviations is therefore symmetrical.

[5]Incidentally, the standard deviations of the Numerical and Verbal parts of our scholastic aptitude test (Figure 4-2) appear by inspection to be approximately 15 in the Numerical distribution and 5 in the Verbal. Look at Figure 4-2 and see whether you agree.

variability indices, it varies least among random samples of the same population. The common statistics to which it lends itself are the product-moment coefficient of correlation and the various standard errors, both of which will be discussed later.

The Semi-interquartile Range

A much easier statistic to comprehend (and to compute) is the *semi-interquartile range*. Look at Figure 4-3. What you see is a negatively skewed distribution cut into four equal parts. At the upper end of each of those quarters is a point on the base line called a *quartile* (Q)—the first quartile (Q_1) above the lowest quarter, the second (Q_2) above the lowest two quarters,[5] and the third (Q_3) above the lowest three quarters. The point just above the highest score in the distribution would logically be Q_4, but that expression is seldom if ever used.

Essence of the concept

The *semi-interquartile range* is simply the mean of two intervals: Q_1 to Q_2 and Q_2 to Q_3. To compute it, you divide the Q_1-to-Q_3 interval by 2.

Appropriate applications

The semi-interquartile range is to measures of variability what the median is to measures of central tendency. Although somewhat less stable, it is preferred to the standard deviation in the same kinds of

Q_1 Q_2 Q_3

FIGURE 4-3. Negatively skewed distribution with area divided into quarters.

---

[5]What other familiar statistic has essentially the same definition? If you are not sure, turn back to page 18.

situations in which the median is preferred to the mean—namely, when distributions are radically asymmetric. It is the perfect companion to the median wherever the latter is properly applied.

The Range

There is another measure of variability—one that probably gets more attention than it deserves, both in makeshift analyses and in this book. It is given attention in makeshift analyses because it is so easy to compute, and much of the space assigned to it in this book is occupied by (1) an exposure of its fundamental weakness and (2) an attempt to prepare you for possible differences in its interpretation.

Essence of the concept

There are some occasions on which we are interested specifically in the most extreme cases in a sample; on those occasions we may report its *total range*—or simply *range*—which is the distance from the lowest score to the highest, plus 1. Each of the extreme scores (probably only one at each end of the distribution) occupies a class interval 1 unit long; so, after we have found the distance from the lowest midpoint to the highest, it will be necessary to add .5 at each end in order to include the entire distance from below the lowest occupied interval to above the highest. In Figure 3-1 (on page 14), the lowest score (midpoint of a class interval of 1) is 2; the highest is 9. Subtracting the lowest from the highest would give us 7; but we really should be subtracting 1.5 from 9.5, which would give us a range of 8. The easiest way to compute the range usually is first to *find the difference between the midpoints* of the two class intervals in which the extreme scores lie and then *add 1*. You may find other textbooks, however, that do not advocate adding the 1. Some authorities regard that refinement as hair splitting, because the range is such an unstable statistic in any case*.

In fact, the most important feature of the range is a weakness—its extreme instability. We have just noted that the range in Figure 3-1 is 8. Now turn to Figure 3-4. It is the same as Figure 3-1, with the exception that two scores have been moved away from the main group. Note the effect on the range. (Instead of 8, it is now $35\frac{1}{2}-1\frac{1}{2}$, or 34!) Note, too, that not even *two* extremely high scores were necessary to have that effect; one would have done exactly the same thing. The fact is that the range is determined by *two, and only two, score intervals* in any distribution: the lowest and the highest. That is why it is so easy to compute; that is also why it is so unstable.

Appropriate applications

We have seen that the standard deviation is analogous to the mean and is its proper companion, and we have noted a similar relationship of the semi-interquartile range to the median. The relationship of the range to the mode is not quite as neat, but there are similarities that should help you to remember both statistics.

One similarity is that each is the *quickest* estimate of its kind. Another, closely related, one is that each is *less stable* than alternative indices. Finally, both the mode and the range are used, more often than are other statistics, to answer questions related directly and very simply to their definitions. For the mode, the question is, "Which is the typical (most frequent) case?" For the range, it is, "How much of the scale must be used to represent the distribution?"

Summary

Two important ways of describing a sample are (1) by its central tendency or average, which indicates the general level of the scores, and (2) by its variability, which tells the extent to which individual scores deviate from that average. This chapter has been about the latter type of description—measures of *variability*.

You have been invited to use your previously developed understanding of measures of central tendency as an anchor for the new concepts; the latter were presented as analogs of the former. Specifically, the *standard deviation* is roughly analogous to the mean and is its companion statistic; the *semi-interquartile range* goes with the median; and the *range* is similar to the mode.

The *standard deviation* is a kind of average of individual deviations from the mean. In a normal distribution, it is the most stable of all measures of variability. It also lends itself to the computation of other stable statistics, like the product-moment correlation coefficient (Chapter Five) and all of the standard errors (Chapter Seven).

The *semi-interquartile range* is the mean of the two middle interquartile ranges, or one half of the distance from Q_1 to Q_3. It is less stable than the standard deviation but is preferred to it whenever the distribution is markedly asymmetric, and it is the perfect companion to the median.

The *range* is the distance from the point just below the lowest score to the point just above the highest. It, like the mode, is easier to compute than any other measure of its kind and, also like the mode, is the least stable among them. It is, of course, the one statistic that gives information directly about the highest and lowest scores in the sample.

We have seen that in a normal distribution, the mean, the median,

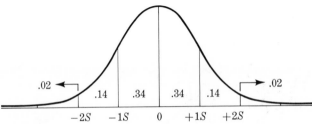

FIGURE 4-4. *Normal distribution with base line divided into standard deviation units.*

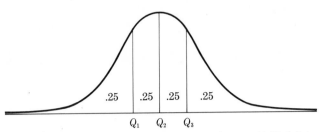

FIGURE 4-5. *Normal distribution with area divided into quarters.*

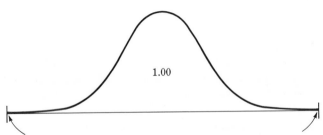

Point just below lowest score   Point just above highest score

FIGURE 4-6. *Normal distribution showing total range on base line.*

and the mode are all at the same place. When we make a similar graphical comparison of measures of variability, the situation is not quite as simple as that.

In Figure 4-4, the base line is divided into equal units—namely standard deviations—and the areas subtended by them are *un*equal. The number above the line indicates the proportion of the total area that is subtended by each segment. The proportions have been rounded off because they are easier to remember in that form, and they are precise enough for our purposes in any case; more exact proportions are given in Figure 6-2.

In Figure 4-5, the base line is divided into unequal segments, and it is the parts of the area that are equal. And of course the range subtends the entire distribution, as in Figure 4-6.

MEASURES OF RELATIONSHIP

There are many times, both in basic science and in professional practice, when we want to know the *relationship* between one thing and another. Indeed, all of science is concerned with such relationships; without knowledge of them, professional practice could never check up on itself.

Take our scholastic aptitude test, for example (pages 12–13). If we merely *assume* that it is doing what we want it to, then much money and even more time and effort may be spent in vain. If, on the other hand, we define the test's effectiveness in terms of its prediction of success in college, then we have a way of checking up on it. We discover how closely the scores are related to some criterion of success, and that tells us how good the test is for that purpose.

What we need, then, is an index of relationship—a number that when low indicates a low degree and when high a high degree of relationship between two variables. We need a *coefficient of correlation*.

Imagine now that the two variables in which we are interested are (1) the height and (2) the weight of a population of toy soldiers. Imagine further that all of those soldiers are exactly the same *shape* and that they differ in size, and therefore in weight.[1] Figure 5-1 pic-

---

[1]This is, of course, an unnatural situation; I am using it because it enables us to focus our attention exclusively on the two variables of our present concern.

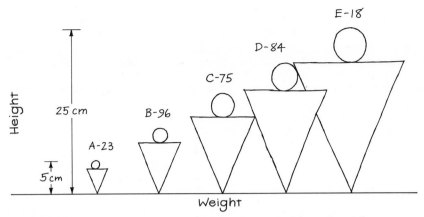

FIGURE 5-1. Sample of population in which height and weight are corre-lated positively.

tures a representative sample of them. If we were actually doing the computation, we should need a much larger sample than five, but for learning the concept, the smaller number is better.

Just by looking at the sample, what would you say the relationship is between height and weight? You notice immediately that the short soldiers are light in weight and that the tall ones are heavy. How would you describe that relationship? Strong? Directly proportional? Perfect? If you were asked to place the strength of the relationship on a scale from .00 to 1.00, you would have to place it at 1.00, because it *is* perfect.*

A coefficient of correlation provides just such a scale—that is, a scale with limits of .00 and 1.00—except that it carries information not only about the *strength* of the relationship but also about its *direction*; some correlations are positive and some are negative.

There are several kinds of coefficients. The easiest to comprehend is the rank-difference coefficient; the most useful (and most used) is the product-moment coefficient. We shall examine each of those in turn.

The Rank-difference Coefficient (ρ)

Near the beginning of the chapter on measures of variability, I introduced you to a statistic that you will probably never see again. I did so because that statistic, the "average deviation," was the best device available for building an understanding of the concept that lies behind the most-used of all measures of variability, the standard deviation. What I am about to do with correlation is not quite as drastic as that, for the *rank-difference* coefficient is used quite often. But it appears in

TABLE 5-1 | *Ordering of Data for Rank-Difference Correlation of Heights and Weights of Toy Soldiers in Figure 5-1*

NAME OR OTHER IDENTIFICATION OF SUBJECT	HIS RANK ON VARIABLE X (HEIGHT)	HIS RANK ON VARIABLE Y (WEIGHT)	DIFFERENCE BETWEEN THE TWO RANKS	SQUARE OF THE DIFFERENCE
E-18	1	1	0	0
D-84	2	2	0	0
C-75	3	3	0	0
B-96	4	4	0	0
A-23	5	5	0	0
Σ				0

the literature much less often than the *product-moment coefficient*, and again I am presenting the less frequently used statistic first because it illustrates more directly the essential nature of the more common one.

Positive correlation

Table 5-1 shows how the data would be ordered in preparation for computing a Spearman *rank-difference* coefficient of correlation on the toy soldiers in Figure 5-1.

In Table 5-1, the first three columns list each soldier's (1) serial number, (2) rank on the variable *height*, and (3) rank on the variable *weight*. The fourth column shows the difference between the two rankings. If you will draw a straight pencil line from $X1$ to $Y1$, another from $X2$ to $Y2$, etc., the lines will form a ladder pattern. (In a moment, we'll look at a similar table that forms a different pattern.) Notice, too, that *all the rank differences are zero.* Keep that in mind while looking at the formula for the ρ (*rho*) *coefficient*:

$$\rho = 1 - \frac{6\Sigma D^2}{N(N^2 - 1)} \tag{5-1}$$

where D is the difference between the two ranks of a single subject and N is the number of subjects—that is, the number of pairs of measurements. We are not interested in the computation as such;[2]

---

[2]If you *should* ever need a quick estimate of correlation, this is the one to use. Consult any standard statistical methods text about what to do with tied ranks, and within a few minutes you'll be ready to use this formula.

the formula is included here only to demonstrate that the rank differences are in the numerator of a fraction that is subtracted from 1.00. That means that the larger the rank differences, the smaller the rho, down to a limit of .00. (After that, still larger differences contribute to rising *negative* coefficients, up to a limit of -1.00.)

Any correlation coefficient carries information about two aspects of a relationship: (1) its *strength*—measured on a scale from zero to unity— and (2) its *direction*—indicated by the presence or absence of a minus sign. Consider again our example in Figure 5-1. I had asked you to imagine that all of the subjects were exactly the same shape, and I attempted to draw them all the same shape. But my draftsmanship falls short of perfection (only slightly, mind you!), so if we were to cast those toy soldiers in molds made directly from my drawings, the relationship between height and weight would *not* be perfect, after all. You might expect a coefficient of correlation to be sensitive to that difference between perfection and near perfection; but the rank-difference coefficient is not. Examine Figure 5-1 again, and you will see that all the subjects clearly are ranked the same even when the imperfections of my drawing are recognized. I mention that here because I want to show you later an index (the product-moment coefficient) that *does* take those imperfections into account.

As for the negative sign that accompanies some coefficients, there is an important point to be made (or rather, emphasized, for it has already been made; see above). *The size* (strength) *of a coefficient is entirely independent of its direction* (positive or negative). Although the formula for the rank-difference coefficient does not encourage it, we should think not of a single continuum from negative 1.00 through .00 to positive 1.00 but rather of two separate dimensions—one negative, one positive—each of which begins at .00 and ends at 1.00. A correlation of 1.00 is no bigger (stronger) than a correlation of -1.00.

Negative correlation

But what does it mean to say that two variables are *negatively* correlated? Let us look again at the variables of height and weight, but this time in a population quite different from the toy soldiers that we examined in Figure 5-1 and Table 5-1. Consider now a human population in which the shortest individuals are the heaviest and the tallest are the lightest. (Such an arrangement is beyond your experience, but hopefully not your imagination.) Figure 5-2 is an attempted rendering of a sample of five drawn from such a population. The corresponding table (5-2) reveals the "different pattern" I promised a moment ago. Now if you draw pencil lines between equivalent ranks, you will find that a star pattern emerges instead of a ladder.

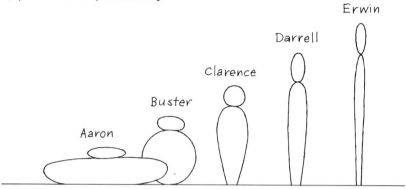

FIGURE 5-2. *Sample of population in which height and weight are correlated negatively.*

And if you sum the squared rank differences, you will find a marked contrast with the zero that comes from Table 5-1. (Remember that D^2 is in the numerator of a fraction that is *subtracted* from 1.00 to obtain the correlation coefficient.) But my drawing is a considerably less adequate representation of a perfect relationship than it was in Figure 5-1. Does the rho coefficient detect my inadequacy this time? No. Nothing will affect it until the *ranks* change. In Table 5-2, the correlation of height to weight is a full −1.00, even though the weights of Darrell and Erwin, for example, are perceptibly farther apart than their heights.[3]

In other words, information is lost when we convert interval measurements into rank orders. That difficulty can be overcome, at considerable cost in computational labor, by using a different coefficient of correlation. The next section will aim at an understanding, without computational labor, of that coefficient.

The Product-moment Coefficient (r)

The rho coefficient is included in this book primarily because it makes a better vehicle for teaching than the more difficult Pearson *product-moment coefficient.* You will see rho reported from time to time in the

---

[3]That difference may not be as perceptible to you as it is to me, so let me give you an exaggerated example. In the following diagram, the arrangement of Subjects A, B and C on Variable X is quite different from that of the same individuals on Variable Y; but their ranks are the same on both.

Variable X	A		B		C
Variable Y	A			B	C

TABLE 5-2 | *Ordering of Data for Rank-Difference Correlation of Heights and Weights of Men in Figure 5-2*

NAME OR OTHER IDENTIFICATION	HIS RANK ON VARIABLE X (HEIGHT)	HIS RANK ON VARIABLE Y (WEIGHT)	DIFFERENCE BETWEEN THE TWO RANKS	SQUARE OF THE DIFFERENCE
Erwin	1	5	−4	16
Darrell	2	4	−2	4
Clarence	3	3	0	0
Buster	4	2	2	4
Aaron	5	1	4	16
Σ				40

literature, but usually it is used only (1) as a relatively quick approximation of the "Pearson r," as the product-moment coefficient is often called, or (2) in situations in which the assumptions for r cannot be satisfied.*

The meaning of "product-moment"

To begin our discussion of this more sophisticated index, let us look again (Figure 5-1) at those toy soldiers. Once again, we construct a table, and indeed it is of the same general form as Table 5-1; but this time (Table 5-3), instead of *ranks*, we enter ordinary interval scores that tell also how far apart the subjects are on each variable. (See the footnote on the preceding page.)

TABLE 5-3 | *Ordering of Data for Product-Moment Correlation of Heights and Weights of Toy Soldiers in Figure 5-1*

SERIAL NUMBER	X†	Y‡	x	y	x^2	y^2	xy
E-18	25	250					
D-84	20	160					
C-75	15	90					
B-96	10	40					
A-23	5	10					

†Height in centimeters.
‡Weight in grams.

TABLE 5-4 | *Effect of Ordering on Sum of Cross Products* (Σxy)

	ORDERED					RANDOM				
X	Y	x	y	xy		X	Y	x	y	xy
25	250	10	140	1400		25	160	10	50	500
20	160	5	50	250		20	40	5	−70	−350
15	90	0	−20	0		15	10	0	−100	0
10	40	−5	−70	350		10	250	−5	140	−700
5	10	−10	−100	1000		5	90	−10	−20	200
Σ 75	550			3000		75	550			−350
$\overline{X}$ 15	110					15	110			

Don't be intimidated by that table. One reason it is so large is that information must be collected for the computation of two standard deviations. (The x^2 and y^2 columns are used exclusively for that.) The reason for including that computation will come later;[4] at the moment, it is sufficient merely to note the presence of S_x and S_y in the formula for r:

$$r = \frac{\Sigma xy}{NS_x S_y} \tag{5-2}$$

The main idea behind that formula is that when X and Y scores are arranged "in parallel," so to speak, the product of the corresponding deviation scores (xy) is maximally large. That is so because in such an arrangement the largest deviation scores—both positive and negative—occur together in the table and thus are multiplied together. Compare the two parts of Table 5-4 and you will see what I mean.

Close examination of the two parts will reveal that the scores are the same in both but that they are arranged differently. On the left, the arrangement is perfectly ordered, as the drawing of toy soldiers on page 31 would suggest; on the right, the arrangement of the Y scores is random, which means that the relationship between X and Y is also random.

Now see what that does to the correlation coefficient. On the left, every large positive deviation score gets multiplied by another that is

---

[4]You will be in a better position to understand that reason after you have been introduced to the concept of "standard scores" in Chapter Six.

similarly large and also positive; whereas on the right, a large positive deviation may be neutralized by a low multiplier, and *any* deviation may be multiplied by one of an opposite sign, thereby yielding a negative product. The result is that with the random arrangement, the sum of the *cross products* (Σxy)—and hence the correlation coefficient—is always small. (In fact, it varies from *zero* only by chance!) In terms of the "weightless beam" concept introduced on page 14, you could say that the deviation scores on the two variables are often either close to the fulcrum or on opposite sides of it; whereas in a more substantial correlation, *x* and *y* scores for a given individual are consistently together—many of them far from the fulcrum—so that they can apply an extremely large torque, or *moment*, to the beam. Hence the term "product-moment"—a reference to "the product of the moments," which is largest when all of the scores are perfectly ordered.

The scatterplot

There is another way of looking at correlation that gets directly at the fundamental idea. I refer to the *scatterplot*. A scatterplot is a surface on which both the *X* and the *Y* scores of each individual can be represented by a single point. If, for example, we were correlating scholastic aptitude test scores with first-year grade-point averages and if every student's test score were precisely proportional to his subsequent scholastic performance, the scatterplot might look like Figure 5-3. Each student is represented by a single point (represented here by a black dot) that locates him on two dimensions, *X* and *Y*. In this case, all of those points lie in a straight line (a line that is called a *regression* line because it embodies the mathematical regression of *Y* on *X*.)[5] The "scatter" that we have been discussing determines the slope of the regression line; the smaller the scatter, the steeper the slope and the bigger the correlation. In our illustration, there is *no* scatter, which is how it always is with a perfect correlation.

Notice, however, that at least some of the students in our example

---

[5]The slope of a line in a coordinate plot like this is the ratio of the amount of change on the ordinate (*Y*) to the amount of change on the abscissa (*X*). If *Y* increases $\frac{1}{2}$ unit for every unit increase in *X*, the slope is .5. If while following a regression line with your pencil you find that you have to move four units on the *X* axis for every one you cover on *Y*, the slope is .25—providing that the changes on both variables are positive. The same *degree* of slope could occur in a negative *direction* if *Y* were *dropping* $\frac{1}{4}$ unit with every one-unit rise in *X*. The relation between scatter and slope is explained later (pages 58ff), after you have mastered the concept of the standard score. For an interesting historical account of the development of regression and correlation concepts, see Guilford, J. P., *Fundamental Statistics in Psychology and Education*, 4th edition. New York: McGraw-Hill, 1965.

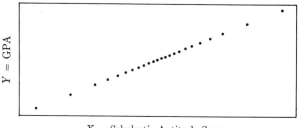

X = Scholastic Aptitude Scores

FIGURE 5-3. *Scatterplot of perfect positive correlation.*

could deviate perceptibly from the regression line without changing any *ranks* on either variable. Select some adjacent individuals (at the upper or lower ends of the line, because they are farther apart there) and try it a few times with your pencil. You will find that within the "same rank" limitation, there is considerable freedom to change measured values on X and Y. The amount of such freedom is the amount of information that is lost when interval scores are converted to ranks. That information is retained when the correlational statistic is a Pearson r.

If, as in real life, the correlation is *not* perfect, there is always *some* scatter. But if the X-Y relationship is extremely strong, the deviations are small (see Figure 5-4); we can make very accurate predictions of Y scores from a knowledge of X scores. At the other extreme is the "random" X-Y relationship depicted in Figure 5-5. There, there is no tendency for a low Y to be associated with a low X or a high Y with a high X, and as you can see, there is a near maximum of scatter. Knowing a person's score on X does not help at all in estimating his score on Y.

Negative correlations behave in exactly the same way, except that the regression line slopes *down* from left to right instead of up. Figures 5-6 and 5-7 are examples.

Figures 5-3 through 5-7 have been contrived to represent perfect positive, high positive, zero, high negative, and perfect negative correlations. Either of the "perfect" correlations would enable us to predict Y precisely from a knowledge of X, and vice versa. The zero correlation would tell us that a knowledge of an individual's position on one variable would be useless in predicting where he is on the other. Such a correlation would say that high scorers on the scholastic aptitude test are just as likely to flunk out of school as low scorers. The other correlations (high positive and high negative) would not insure that we could make *perfect* predictions, but they would significantly *reduce our errors*, should we wish to try predicting college performance from test scores.

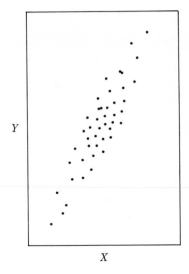

FIGURE 5-4. *Scatterplot of high positive correlation.*

Y

X

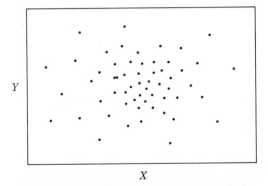

Y

X

FIGURE 5-5. *Scatterplot of zero correlation.*

Figure 5-8 depicts those relationships in a slightly different way. Of the four lines that cross diagrams A, B, and C, one line (the perpendicular) is the mean of all the X scores, another line (the horizontal) is the mean of the Y's, and the third and fourth lines (the diagonals) are regression lines. The reason that you don't see four lines at $r = 1.00$ (diagram A) and $r = .00$ (diagram C) is that when one line corresponds exactly to another, you can see only one. In diagram A, where the correlation is perfect, the regressions of y on x and of x on y are merged into one line that is just 45° from each of the two base lines. In diagram C, where $r = .00$, the regression lines are separate, but the regression of Y on X is at the mean of Y all the way across, and that of X on Y is

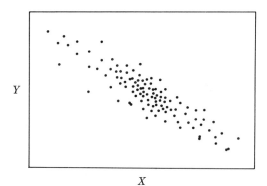

FIGURE 5-6. *Scatterplot of high negative correlation.*

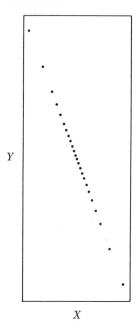

FIGURE 5-7. *Scatterplot of perfect negative correlation.*

at the mean of X; so the two regression lines are superimposed on the two mean lines. (They *are* the two mean lines.) Thus, you see two lines instead of four.

Have you noticed that in contrast to Figures 5-3 through 5-7, all of the scatterplot grids in Figure 5-8 are precisely square? There is a very good reason for that, as you will see later; but it cannot be explained

FIGURE 5-8. *Three scatterplots on standard grids (all scores in standard units).*

A

B

C

FIGURE 5-9. *Three-dimensional rendering of two scatterplots.*

until you have mastered the concept of the "standard score." Watch for that concept when you read Chapter Six, and after you've mastered it return to get another perspective on Figures 5-3 through 5-8.

There is one feature of Figure 5-8 that could be misleading if not accompanied by some explanation. It is especially noticeable in diagram A in which it appears that there is, after all, some scatter when the correlation is perfect ($r = +1.00$).

That is an erroneous impression. What has happened is that dots that should have been stacked directly on top of each other were dispersed horizontally so you could see them. Figure 5-9 is an attempt to represent three dimensions. You can see there that when the correlation is perfect, there is *no* scatter. In stark contrast, a zero correlation spreads out over most of the board. Other correlations approximate those two in varying degrees.

Effect of Restricted Variability

Everything I have said about correlation is based upon the assumption that the individuals measured constitute random samples from both distributions. If the correlation were between IQ and scholastic achievement, for example, we should select a large sample of subjects and measure each of them twice—once for IQ and once for achievement. Their scores on the IQ test should form a distribution similar in every way but size (N) to the one that would have emerged had we tested an entire *population* in the same way; and similarly for achievement.

On the other hand, we saw very early (pages 12–13) that it is sometimes useful to subdivide a general population into subcategories and that it is legitimate to call each *sub*category a "population," too.

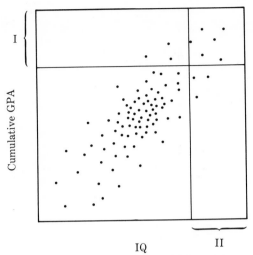

FIGURE 5-10. Scatterplot of high positive correlation, showing effect of restricted variability.

For example, instead of including literally *everybody* in our population of IQ's, we could arbitrarily define a population that includes only college students. *Now* our "sample" must be taken at random* from *that* "population."

All of this is fine as far as it goes, but we must recognize that it has a very significant effect on the correlation coefficient. The best way to demonstrate that fact is again to draw a diagram; Figure 5-10 is appropriate to our present need. It shows a hypothetical relationship between intelligence (as measured by a standardized test) and achievement (as measured by grade-point average) in the general population. It also shows what the relationship would be if the population were defined as "applicants to graduate schools." The entire range of IQ scores for the general population forms the base line of the diagram, and the entire range of grade-point averages forms the ordinate. Brackets I and II enclose those segments on each scale within which 90 percent of applicants to graduate school may be found.

The scatterplot in Figure 5-10 reveals a substantial correlation between intelligence and scholastic achievement. However—and this is the point toward which this discussion has been directed—a plot of the area marked off by segments I and II reveals no relationship at all. Figure 5-11 is an enlargement of that area.

Be cautious, then, whenever you interpret a low correlation coefficient. In the above illustration, it would be proper to say that among

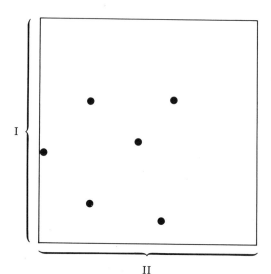

I

II

FIGURE 5-11. *Restricted area of scatterplot (enlarged).*

applicants to graduate schools, there is virtually no relationship between IQ and GPA; but be careful *not* to infer from that *a similar lack of relationship in the general population.* A correlation coefficient is relevant only to the population from which the sample has been drawn.

Correlation and Causality

Before we can be certain that we have found an instance of *causation*, we must observe a reliable relationship between the purported causative agent and the event being caused. If every time you push a switch in your bedroom to the "up" position a light goes on, you come to believe that the latter event is caused by the former, even in the absence of wiring diagrams, electrical theory, and so forth. If your dog becomes wildly aggressive whenever the garbage man approaches, you say that the man's approach is causing the dog to be upset. If in studying a large sample of school children you find that those with larger vocabularies have higher IQs, you assume that the vocabulary is a cause of the intelligence.

But should you? Is it legitimate for you to assume that when two events occur together one is caused by the other? Consider the following case: what size of coefficient would emerge if we were to compute a correlation between the revolutions per minute of the left and right

Left

Right

FIGURE 5-12. Schematic automobile chassis.

front wheels of your car (Figure 5-12)? The correlation would be pretty high, would it not? If the car were on a curving road and if turns in one direction were predominant throughout any given minute, the two scores would not be the same for that minute; but with varying speeds on a straight road, the correlation would be almost perfect (Table 5-5).

TABLE 5-5 | *Arrangement of Data in Correlation*
of RPMs Over Minutes of Travel

SUCCESSIVE MINUTES OF TRAVEL	REVOLUTIONS OF LEFT WHEEL	REVOLUTIONS OF RIGHT WHEEL
1	?	?
2	?	?
3	?	?
. . .	. . .	. . .

But does the speed of one wheel *cause* that of the other? Only to the extent that each is a functioning part of a total system in which the two events occur. It would be more accurate to say that the functioning system—the car in motion—causes the revolutions of *both* wheels.

I don't suppose anyone will ever bother to find the correlation between the two front wheels of his automobile, but the same principles apply also to many situations in which correlation coefficients *are* commonly computed. For example, if we were to study a large sample of elementary school children, we might find a tendency for those children who display extensive vocabularies to be superior students in arithmetic. We might be tempted to conclude that a good vocabulary causes competence in arithmetic. But then we find that chronological age correlates with both variables. Might it not be better to say that attainments in both vocabulary and arithmetic are "caused by" (are parts of the functioning system of) the child's developing intelligence?

The caution that I am advocating here has important practical implications; for without it we might easily be persuaded, for example, to set up elaborate vocabulary-training programs for children who show weakness in quantitative thinking. Some wag once noted that there is a substantial correlation between the intelligence of boys (as indicated by their mental ages) and the length of their trousers. He suggested that a relatively inexpensive way to increase the intelligence of boys would be to increase the length of their trousers!

So although correlation is a *necessary feature* of a causal relation, it is not a *sufficient proof* that a causal relation exists. Whether the relation is interpreted as a causal one should depend not just on the correlation of two variables but also on some rational link between them—on the extent to which the relationship "makes sense" within some sort of conceptual framework (within a wiring diagram, for example, or a sociological theory).

Correlation is not causation.

Reliability and Validity

At the beginning of this chapter, I suggested that an index of correlation would allow us to check on the accuracy of predictions made by a scholastic aptitude test. Now, by substituting the test score for X and college GPA for Y in each of the diagrams in Figure 5-8, you can see just how accurate the predictions would be if the correlation were .00, if it were .75, and if it were 1.00.

A correlation coefficient that is used, as in the case of the above-mentioned scholastic aptitude test, to predict a *criterion* score (Y) from a test score (X) is called a *validity coefficient*.[6] (The *criterion* is the entity that our test is supposed to be measuring—in this case, "scholastic aptitude," or the capacity to profit from schooling.[7]) If we were to administer our scholastic aptitude test to all entering freshmen as indicated earlier, and if we were to test them again a week later, we

---

[6]The examples given are of "predictive validity" and "test-retest" reliability. There are other kinds of validity and of reliability, but these examples serve to illustrate how a correlation coefficient *can* serve as an index of either.

[7]GPA is generally accepted (though reluctantly by some) as an index of scholastic achievement—i.e., the extent to which an individual *has* "profited from schooling."

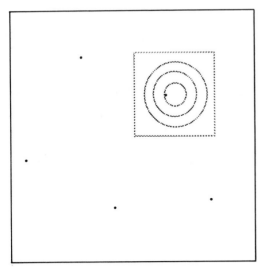

FIGURE 5-13.   Sheet of white gauze with invisible target; shots widely spaced.

could compute an *r* that would tell us the relationship between the first and second testings. Instead of an r_{xy}, we would have an r_{xx}, and it would be called a *reliability* coefficient, because it would tell us the extent to which we can depend on the test to give us the same results from one administration to the next. (You can't measure anything reliably with a rubber yardstick!)

The "rubber yardstick" metaphor is a good one to help you remember the concept of reliability, but it bears no direct relation to validity. Here is an analogy that illustrates both reliability and validity:

Imagine a man aiming a rifle in the general direction of a target that is several hundred feet away from him. I say "in the general direction" because there is a huge sheet of white gauze that hides not only the target but a considerable area around it. Our marksman takes aim, then, at a spot that he hopes is the bull's eye. He fires five shots—and scatters them all over that cloth screen. Figure 5-13 shows the pattern that they make on the screen and on the hidden target.

Another marksman takes aim at the same spot as the first one did. He "groups" his shots as shown in Figure 5-14, thereby missing the target completely.

And finally, a third marksman decides where *he* thinks the target is, aims at its supposed center and pumps off five shots that are as close together as those of the second marksman and are also perfectly placed on the target as in Figure 5-15.

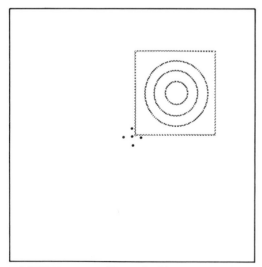

FIGURE 5-14.   Sheet of white gauze with in-
visible target; shots grouped closely together off
target.

Now review the concepts of reliability and validity, and see if you can tell which of them is illustrated by which marksman. Reliability? That would have to be the closely grouped patterns, wouldn't it? Shot after shot lands in nearly the same place; you can count on it, just as you can count on a reliable test to yield nearly the same scores when the same people are retested. Proximity to the target, on the other hand, is the analog of validity. Close grouping of shots doesn't help a bit unless it is on target. Reliable testing is useless unless we know what it is we are testing.

So high reliability does not insure high validity. But it is also important to note that *low* reliability absolutely precludes it. Unless your shots are all very close together, it is impossible to make a perfect score. If a test has low reliability, it can have *some* validity (see Figure 5-13) but not much.

To put it another way, a test can be very reliable without being valid, but it can't be very valid without being reliable.

If when you have finished this book you continue your study of measurement, you will have an opportunity to develop a better understanding of the kinds of validity and reliability that I have barely touched upon; you will also discover that there are others that I have not even mentioned. We will not go further into those matters here, but do be alert to later applications.

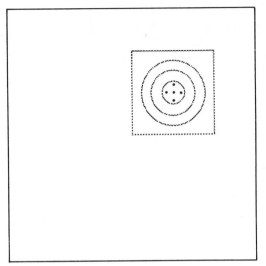

*FIGURE 5-15.   Sheet of white gauze with in-
visible target; shots grouped closely together on
target.*

Summary

It is often important to know the relationship between two variables.
Coefficients of correlation are indices of relationship. Many such indices
have been devised for various applications, but we have discussed
just two: the rank-difference and the product-moment coefficients.

Any *correlation coefficient* is an index of the extent to which measurements of the same individuals are to be found on corresponding
segments (low, middle, high) of two different scales. When such a relationship holds precisely and with no exceptions, the correlation is said
to be perfect, and the coefficient is 1.00. Less than perfect relationships
produce coefficients smaller than 1.00. A *negative* correlation is just as
strong as a *positive* correlation if its coefficient is equally large; the
coefficient indexes magnitude and direction separately in a single number that (1) varies from .00 to 1.00 and (2) does or does not carry a
minus sign.

The Spearman *rank-difference (rho)* procedure places all subjects in
order from smallest to largest on the basis of their scores on X; then it
ascertains how nearly their Y scores approximate that order. The better the approximation, the smaller are the differences in rank (hence
the name "rank difference") and the larger the coefficient of correlation.

The Pearson *product-moment coefficient* (r) is like rho except that it retains information that rho throws away—namely, the distances that separate subjects of adjacent ranks on either of the variables being correlated.

If the position of every subject is plotted in terms of both X and Y variables, a pattern called a *scatterplot* emerges. In that pattern, it is possible to perceive directly the degree of relationship between the two variables.[8] The same diagram reveals the direction of that relationship.

Correlation procedures have many uses in the technology of testing, but two are particularly noteworthy. In the kinds of situations cited here, the *coefficient of reliability* tells us whether a test does *consistently* whatever it does; the *coefficient of validity* tells us whether it is doing what we have *believed* it to be doing.

The formula for r deals primarily with the products of individuals' deviation scores, but an important part of it is also the two standard deviations in the denominator. An adequate explanation of their presence there awaits your understanding of the "standard score"—a concept that is the core of our next chapter.

---

[8]We might still want to compute r, however; the most *direct* index of an entity is often not the most *precise* one.

INTERPRETING INDIVIDUAL MEASURES

Measurement has been defined as "rules for assigning numbers to objects to represent quantities of attributes."[1] The importance of *rules* lies primarily in the fact that the resulting "assignment of numbers" nearly always functions as a *communication*, and unless the recipient of a communication knows the rules by which the sender has made that assignment, he will be unsure of its meaning. Many such rules are intuitively obvious. If you tell me that you have measured the distance from Point *A* to Point *B*, I shall assume that the rule you followed was to lay a ruler across the two points and note how many units (for example, inches) lie between them.

But most rules are not so obvious. What rule do you follow if you want to tell me the size of a particular circle? Do you lay your ruler across the center and the circumference and report the number of units

[1]Nunnally, Jum C., *Psychometric Theory*. New York: McGraw-Hill, 1967, p. 2. [Copyright 1967 McGraw-Hill. Used with permission of McGraw-Hill Book Company.]

between the two? If so, you had better label the resulting number as a "radius," because then I shall know exactly what you did to get it. There are, of course, two other labels—"diameter" and "circumference" —that you could use and that would specify two rather different operations.

There are alternative operations for the circle, but they have become standardized over the years so that their descriptions are no longer necessary to precise communication; mere labels suffice. In some educational and psychological measurements, a similar standardization has occurred—though it is not as complete because the operations are much more complex. "Level of intelligence," for example, can be represented by a number; but the number is rendered more meaningful by a specification of the particular standardized test from which it came; and even then, the specificity of the operations falls far short of that of the simple physical examples described in the previous paragraph.

Other measurements are not standardized at all. If we want to do a rat study in which one of the variables is "drive level," we have to figure out a way to measure that level; then, having done so and run our experiment, we are obliged to describe in some detail the operations that resulted in whatever scores (measures) we have to report. It is in situations like this one that the need for specifying rules is most apparent; but the rules are there even (especially?) when, as in the measurement of a circle, they are not explicitly stated. Were it not so, there could be no communication.

Our definition says that the rules we have been discussing are "for assigning numbers to objects . . ." It might be worth a couple of sentences to point out that numbers are not always so used. A mathematical system (a set of rules for the manipulation of symbols) can make use of numbers without any reference to objects. Such a system is viable to the extent that (1) its operations are internally consistent and (2) there is no requirement that it bear any relation to the real world.

So numbers are not always assigned to objects. In measurement, they are so assigned. But even in measurement, a number does not represent an object as such, but rather, the quantity of some *attribute* of the object. How smart is this boy? How tall is that toy soldier? How long is the segment *A–B* on the surface of this burial site? The numbers that we assign to the objects represent the magnitude of their attributes.

The definition of measurement that was quoted earlier (page 51) is adequate as far as it goes; it is a good general definition. But for our particular purpose, one further point needs to be made: in many measurement problems, scores can legitimately be interpreted only as *individual differences.* Even in a notably sequential discipline like mathematics, when your teacher tells you that he is grading his students "on absolute standards," there is reason to suspect that his statement is not entirely correct. He may require that first-year work be mastered before he "passes" you to the second year; but he must have some idea, either from his own experience or that of his mentors and colleagues, what can reasonably be asked of first-year students. His "absolute standards" turn out to be less absolute than he thought.

The example is from education. I am not saying that it is never possible to make any educational measurement without taking into account many other measurements on the same dimension. What I am saying is that in most applications, the more those other measurements are utilized in the interpretation of an individual score, the more sophisticated the interpretation will be.

Consider the case of Joseph O. Cawledge. JOC's high school grades were pretty bad; but then he was a full-time athlete, and the rest of the school's program never interested him very much. How well might he perform scholastically if he were motivated to do so? To find out, we give him Cerebral University's scholastic aptitude test.

You will recall that the CU test is divided into two parts—a "Numerical" test and a "Verbal" test. Joe Cawledge takes them both because they are both required of all entering freshmen. He gets a score of 60 on the Numerical test and a score of 30 on the Verbal. Now, what do we know about Joe?

It *looks* as though he is twice as good with numbers as he is with words. But mightn't that appearance be deceiving? Turn to page 24 and place each of Joe's two scores within the appropriate distribution in Figure 4-2. (Insert a bookmark there, for we shall be using Figure 4-2 repeatedly for the next few moments. In these as in any interval measurements, two questions must be answered about each scale: (1) what is its *point of reference* and (2) what is its *unit of measurement*? Answers to those two questions about a scale would make it possible for us to answer two parallel questions about any individual score: (1) is it *above or below* the reference point and (2) *how far* is it from that point?

It is customary to use the mean of some well-defined "standardization" sample as the standard reference point. You can see immediately that each of Joe's scores is above that point on the appropriate scale. But *how far* above? "Ten points," you say? "They are the *same* distance above their respective means." Look again. One score (N) is in the populous middle part of its distribution, and the other (V) is way out in the "tail" of its distribution. On second thought, then, doesn't it seem to you that the latter must be a much higher score than the former?

Yes, but *how much* higher? Once again we find that drawings are excellent aids to a comprehension of basic principles, but for precision we need a numerical index. We need a single number that will tell us how far away from the mean any given score is and in what direction.

Standard Scores: The z Scale

The answer to the question "how far?" must be in terms that make it possible to *compare* an individual's score on two different scales. To put it another way, we must find a way to put both scores onto a single scale—a *standard scale*, if you will. That probably doesn't mean much to you right now, but in a moment it will.

We have already found a common reference point for both distributions, have we not? Every distribution has a mean; therefore the mean can be used as the reference point for the standard scale that we are about to develop. Every score that falls precisely on the mean of its own distribution will be given a score of zero, because all counting of units will start from there.

Now that we have a common reference point, all that remains is to find a common *unit* for our new scale. As Figure 4-2 demonstrates, it must be a unit that can be used to *compensate for the variability* of a distribution; and in order to do that, it must *be* a measure of variability. If we were to divide Joe's Numerical deviation score by a large number and his Verbal score by a small one (look back at Figure 4-2), wouldn't we have a realistic index of his position in each distribution?

A measure of variability *would* be large for the Numerical and small for the Verbal distributions, so that is how we shall accomplish our purpose of converting both scores to a common scale. If you had a measurement stated originally in inches and you wanted to convert it to feet, how would you proceed? You would divide by the number of inches in a foot, would you not? Well, you do the same thing here: you divide by the number of raw-score units in a *standard deviation* (the standard deviation of that particular distribution). The formula looks like this:

$$z = \frac{x}{S} \tag{6-1}$$

where z is a standard score, x is a deviation score, and S is the standard deviation of the distribution in which x occurs.

Let's see how it works in Joe's case. If the standard deviation of the Numerical distribution is 15, and Joe's score of 60 is ten raw-score points above the mean, his deviation score is 10 and his standard score is 10/15, or *0.67*. If the standard deviation of the Verbal distribution is 5, and Joe's score of 30 is again ten points above the mean, his devia-

FIGURE 6-1.   Distribution of standard scores from two different tests.

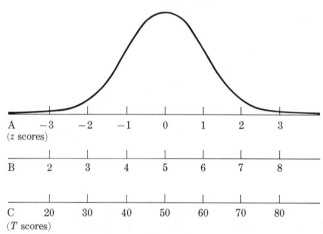

FIGURE 6-2.   Evolution of a standard scale with a mean of 50 and a standard deviation of 10.

tion score is 10 and his standard score is 10/5, or *2.00*. Quite a difference, is it not? Figure 6-1 shows where Joe's two scores would be in a distribution of standard scores from both tests.

We have accomplished our purpose: we have moved scores from two different scales onto a single "standard" scale. When each was expressed in terms of its own scale, the two could not be compared; now they can be.

Other Standard Scores

Effective though it is, our standard scale[2] still has some shortcomings in actual practice. For one thing, the *negative numbers* are a nuisance; for another, the *unit* is awkwardly large.

––––––––––––

[2]Standard scores are sometimes called "scaled scores."

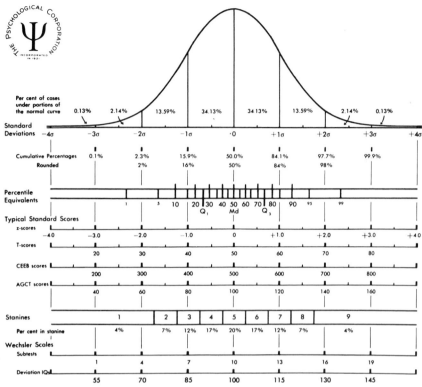

FIGURE 6-3. "The Normal Curve, Percentiles, and Standard Scores. *Distributions on many standardized educational and psychological tests approximate the form of the* normal curve *shown at the top of this chart. Below it are some of the systems that have been developed to facilitate the interpretation of scores by converting them into numbers which indicate the examinee's relative status in a group.*

"*The zero (0) at the center of the baseline shows the location of the mean (average) raw score on a test, and the symbol σ (sigma) marks off the scale of raw scores in* standard deviation *units.*

"*Cumulative percentages are the basis of the* percentile equivalent *scale.*

"*Several systems are based on the standard deviation unit. Among these* standard score *scales, the z score, the T score, and the stanine are general systems which have been applied to a variety of tests. The others are special variants used in connection with tests of the* College Entrance Examination Board, *the* World War II *Army General Classification Test, and the* Wechsler Intelligence Scales.

"*Tables of* norms, *whether in percentile or standard score form, have meaning only with reference to a specified test applied to a specified population. The chart does not permit one to conclude, for instance, that a percentile rank of 84 on one test necessarily is equivalent to a z score of +1.0 on another; this is true only when each test yields essentially a normal distribution of scores and when both scales are based on identical or very similar groups of people.*

"*The scales on this chart are discussed in greater detail in* Test Service Bulletin No. 48, *which also includes the chart itself . . . Copies of this bulletin are available from The Psychological Corporation, 304 East 45th Street, New York, N.Y. 10017*" [*Psychological Corporation, "Methods of Expressing Test Scores,"* Test Service Bulletin No. 48, September 1954.]

Fortunately, however, neither of those shortcomings is difficult to overcome. To get rid of the negative numbers, we have but to add a constant to every score; if that constant were 5, the mean score would be 5, 1S below the mean would be 4, two above it would be 7, and so on. (See line B, Figure 6-2.) To reduce the *size* of the scale unit requires nothing more than *dividing* it into smaller parts, which of course makes the *number* of parts larger. If we want our new units to be one tenth the size of the original, there will be 10 of the new units to every standard deviation. Combining those maneuvers (first adding 5 to every z score and then multiplying the result by 10) produces the scale that you see in line C of Figure 6-2 and opposite the label "T scores" in Figure 6-3.*

Other scales are constructed in much the same way. In the "College Boards" scale (CEEB in Figure 6-3), 5 is added but 100 is the multiplier; the resulting distribution has a mean of 500 and a standard deviation of 100. The Army General Classification Test mean is arbitrarily set at 100 and its standard deviation at 20. In every case, what we have is but a modification of what is still basically a standard-deviation scale.[3]

Standard Scores in Correlation

In a footnote on page 36, I promised that once you had been introduced to the concept of the standard score, you would be ready to develop a deeper understanding of the Pearson product-moment coefficient of correlation. You have now been properly introduced, so let's take another look at correlation.

The reason for the standard deviations that you see in Formula 5-2 is that they are necessary to the conversion of raw scores into standard scores. We may rewrite the formula

$$r = \frac{\Sigma xy}{N S_x S_y} \qquad (5\text{-}2)$$

to emphasize that conversion:

$$r = \frac{\Sigma z_x z_y}{N} \qquad (6\text{-}2)$$

If you remember that z scores are in standard-deviation units (pages 54–55) and also that Σ/N is the way you compute any mean (pages

---

[3]You can afford to ignore the "Stanines" and "Wechsler Scale Subtests," but it would be well worth your while to study the rest of Figure 6-3 until you understand thoroughly all of the conversions that are illustrated there.

13–15), it should be clear to you that r is the mean of the cross products of x and y when *both are expressed in standard units*. It is also *the slope of the regression* of y on x, again when both are in standard units.

If we make scatterplots of ten raw-score relationships that are all of the same strength and direction, we may get ten different slopes of our regression lines; if we first convert all measurements to standard scores, we get identical slopes. Figures 5-3 and 5-7 (pages 38 and 40) are positive and negative, respectively, but they are both perfect relationships; if they were plotted in standard units, the degree (though of course not the direction) of their regression slopes would be the same. Without the conversion, they are very different. Figure 5-8A, also shows two perfect relationships but this time with identical slopes for both; you see only one line. That's because only standard scores are used there.[4]

If your plot is done with raw scores, the length of each unit on either dimension is arbitrary. For example, you may use inches or feet on one dimension, ounces or pounds on the other, and you may let one inch, foot, ounce or pound be represented by whatever distance is convenient, depending on the kind of graphing paper you are using and the horizontal and vertical space that is available. Once the scores have been converted, that is no longer true; for a given strength of relationship, there is one and only one degree of slope. When no relationship exists, the slope is zero; since there is no reason to expect the z_y scores accompanying one z_x value to be any different from those associated with any other, the best estimated central tendency of the z_y's at any given value of z_x is simply the mean of *all* the z_y's in the entire distribution. So as our regression line moves from left to right on the X axis, it moves not at all on the Y; which is to say, the slope of the regression line is zero, as in Figure 5-8C. For a perfect relationship, it is 1.00, as in Figure 5-8A. All other relationships are somewhere between those two extremes.

In short, one meaningful interpretation of r is as *the slope of the regression of* y *on* x *when both are laid out in standard units.*

Centile (or "Percentile") Scores

Standard, or scaled, scores can be used to compare an individual's scores on two different tests or to trace with one test his progress over a period of time. Whatever else it may do, every such score compares

---

[4]If you will turn Figure 5-8 over on its side and examine the regression of x on y, you will discover that in every diagram it is exactly the same slope as that of y on x. (The direction of the slope is reversed because the sequence of y scores is reversed when the diagram is viewed from the side.)

the individual's performance to that of a *standardization or "norm" group*. But the size of that group is commonly monstrous and its composition heterogeneous. Often what is most needed is a comparison with a group that is less cumbersome and more precisely defined.

Take Joe Cawledge's Numerical performance, for example (pages 53–54). Imagine now that the test, far from being homemade by Cerebral University as I have asked you to think of it until now, is only being borrowed by the University as a part of a national testing program. The test has already been *standardized*—that is, it has been administered to a very large number of college students, measures of central tendency and variability have been computed, scaled scores (say T scores) have been derived, and *norm tables* have been prepared that make it possible to convert quickly any individual raw score to other kinds of scores. (You will recall that a raw score by itself is meaningless.) Imagine further that Joe's score on the Numerical test is one standard deviation above the national mean.

That is important information in itself, and it makes possible a comparison of Joe's present performance with his performance on the same test months or years later. What it does *not* do is to compare him with the competition at CU; more specifically, it does not compare his performance with the performances of other entering freshmen at Cerebral University. But the national testing program includes the gathering each year of data concerning those performances—data that show up in next year's norm tables in the form of *centiles*—or, as they are more often called, *percentiles*. Actually, of course, the data themselves are raw scores; some kind of transformation must be accomplished in order for the information to be used in this new way.

What is this new way? Actually, it is very simple. Do you remember the concept of the "quartile" (page 26)? It is the point on the base line that lies just above a specified *quarter* of a distribution. If the point above a quarter is a quartile, how might we refer to the point above a *tenth* of the scores in a sample? *Decile* would be the parallel term. And what if we were to divide the sample into 100 equal parts? The parallel term there would be *centile*, would it not? A centile is the point on the base line that lies above a specified number of *hundredths* of the distribution. The term *percentile* is technical slang for "centile."

Now back to Joe. His Numerical raw score was 60, which happens by coincidence to be precisely one standard deviation above the national mean and converts to a T score that is also 60. That conversion has been done by a computer at national headquarters, and neither Joe nor his counselor ever sees the raw score. What they do see is a standard score of 60 and a *norms table* that allows them to convert it into a "percentile" in any of several populations. (Samples from those populations are called *norm groups*, and the process of gathering data from

TABLE 6-1 | *Norms for Numerical Test*

STANDARD SCORE	PERCENTILES				
	ENTERING FRESHMAN	FRESHMAN	SOPHOMORE	JUNIOR	SENIOR
88					
86					
84					
82					
80					99
78				99	98
76			99	98	97
74		99	98	97	95
72	99	98	97	95	92
70	98	96	95	92	88
68	96	94	92	88	83
66	95	92	88	84	78
64	92	88	84	78	71
62	88	84	78	71	63
60	84	79	71	63	55
58	79	72	63	55	47
56	73	64	56	48	39
54	66	57	48	39	31
52	58	49	40	31	24
50	50	41	32	24	18
48	42	33	25	18	13
46	34	26	19	13	9
44	27	20	13	9	6
42	21	14	10	6	4
40	16	10	7	4	2
38	12	7	4	2	1
36	8	5	3	1	
34	5	3	2		
32	4	2	1		
30	2	1			
28	1				
26					
24					
22					
20					

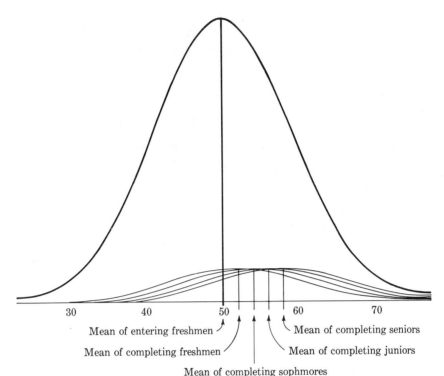

Mean of entering freshmen

Mean of completing freshmen

Mean of completing sophmores

Mean of completing seniors

Mean of completing juniors

FIGURE 6-4.   Distribution of four survivor groups artificially equated in size and superimposed upon original freshman distribution.

them is often referred to as the *standardization* of the test.) Table 6-1 shows how the norms might look for the Numerical test. The table is represented graphically in Figure 6-4.

In all of this, the important thing to notice is that Joe has not one score, but many. If it is true that his performance is meaningless until it has been compared to other performances, it is also true that its meaning is enhanced by a precise definition of the group to which he is being compared at any given time. Once that definition has been made, individuals who fit it can be tested, and the distribution of their scores divided into hundredths. The result is a set of *norms* like that in Table 6-1; now our Joe can see how his performance compares not only to the entering freshmen on whom the Numerical test was originally standardized, but to any other group that has been tested. His score is better than 84 percent of entering freshmen, 79 percent of students completing their freshman year, 71 percent of completing sophomores, 63 percent of completing juniors and 55 percent of graduating seniors. Those norm groups are students from all kinds of programs, including art, music, literature, and so forth, in which little numerical work is

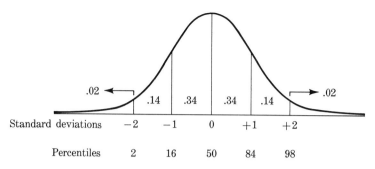

FIGURE 6-5. *Normal distribution with base line divided into standard deviation units, showing proportion of total area subtended by each standard deviation, and listing cumulative percentages (percentiles).*

done. If Joe were thinking of entering an engineering curriculum, it would be appropriate to use norms derived from the testing of engineering students. Meanwhile, we should suspect that when his score is compared to engineering freshmen, it might be well below the median and that among graduating seniors it might even be near the bottom. The small curves in Figure 6-4 have been artificially equated in size and shape. You might expect the "completing freshman" group to be closer to the size of the "entering freshman" group than the drawing would lead you to believe, and you would be right. You might also expect that the *shapes* of the distributions would change systematically from entering freshman to graduating senior; after all, it is the less able whose elimination accounts for the rising averages. But there you would be wrong. In practice, it turns out that distributions of more "select" groups tend to be very nearly normal.

One thing more. In Chapter Four (Figure 4-4) we identified the proportions of the total sample that, in a normal distribution, are subtended by successive standard-deviation units marked off from the mean. Now, percentiles might be thought of as "cumulative proportions." Figure 6-5 reproduces 4-4 and adds the percentile that corresponds to each of the standard scores.

You should be able to reproduce this diagram from memory. But when I say "from memory," I do *not* mean *rote* memory. The only numbers you must memorize are "34 "and "14." Once you have placed those properly, the rest are determined. Try it.

Age and Grade Norms

A percentile tells us where an individual is in relation to a specified norm group. An alternative way to interpret a scaled score (or sometimes even a raw score) is to find that group which the individual's score most nearly resembles. More specifically, each comparison group can be taken entirely from a particular chronological age group or from a particular grade in school.

By far the best known of age norms is the *mental age* (MA) that used to be a prerequisite to the derivation of an intelligence quotient (IQ). An individual child's score is compared to the average* scores of many chronological age groups; the age of the group whose average is closest to his determines his mental age. For example, if a 6-year-old does as well on an intelligence test as the *average* 6-year-old, his mental age is 6;* but if he does as well as the average *8*-year-old does, his "MA" is *8*, not 6.*

It has been more than a decade since standard (scaled) scores virtually took over the field of intelligence testing.* Now, the "comparison group" kind of interpretation is more often manifested in *scholastic achievement* testing as *grade*, rather than age, norms. If a child's score is nearest to that made by the average entering 6th grader, his *grade level* is said to be 6.1, or "the first month of the 6th grade." If the child's score were equal to the average pupil in the middle of his sixth year of school, his grade level would be 6.5. If his performance were most similar to the average seventh-grader in his third month, his grade level would be 7.3; and so on.

Summary

Measurements in the social sciences are nearly always of *individual differences*. The report of a score can seldom be understood out of context; that is, it cannot be understood except in relation to *other* scores.

A *standard score* results from the remarkable feat of placing a score from any test onto a scale common to all tests. It is accomplished by dividing each individual deviation score by the standard deviation of its own distribution.

A *centile* (or *percentile*) shows where an individual stands in relation to a specified norm group. It does so by reporting just what percentage of that distribution falls below his score.

An *age* or *grade* level consists of the identification of the norm group whose average score is most similar to that of the individual being tested.

The essence of each of those procedures is *comparison*—the comparison of an individual to a more or less well-defined group. A standard scale divides the base line of a distribution into equal parts, and percentiles divide the area under the curve (the total sample) into equal parts. Age or grade norms set their points of reference at the average scores of various age or norm groups. So all of these measures are different; but in every one, a new subject's score is compared to many others that have preceded it.

SEVEN

PRECISION OF MEASUREMENT

When Joe Cawledge makes a Verbal test score of 30, we report his raw score, his standard score or one or more percentiles. If someone should want to know the mean score of a group of sophomores in a psychology experiment, we should have the same options there. Of course, neither *raw* score by itself would have any meaning; but what about the other two (standard score and percentile)? You learned in Chapter Six to interpret *them*, did you not?

Well, yes and no. Way back in Chapter Three we discussed briefly (pages 12–13) the problem of "generalizing from a sample"—the question of whether we are justified, once we have measured a sample, in making statements about some population that we say the sample represents. Really, though, that is not an "either-or" question, and a proper answer would have to be phrased in terms of the "extent to which we are justified" in making such statements or of the "level of confidence" we have in making them. The probable *error* of such an inference is often as important to know as the inference itself.

This chapter then, is all about errors. Its title is not misleading, however, for it is by discovering the probable limits of error that we define the *precision* of our inferences. In many applications, that definition is actually more important than the score.

Standard Errors

A *statistic* is a statement about a sample. A mean is a statistic; so is a standard deviation. Together, they describe the sample.

But we are seldom interested primarily in the description of a sample. Usually, what we really want is a description of the *population* from which the sample has been drawn. Unfortunately, it is hardly ever possible to describe the population directly. (If it were, the sample and its population would be one and the same.) What we have to do is *infer* the characteristics of the population from those of the sample. The characteristics of the population are known as *parameters*. "Statistic" and "parameter" are parallel terms, one referring to the sample, the other to the population.[1] If I compute the mean height of 100 college men, I have a statistic; the corresponding parameter is the mean of *all* college men, most of whom I have not measured at all.

Example: the standard error of the mean $(SE_{\bar{x}})$

Let's take another example and work it through in more detail.[2] Suppose we want to know the mean IQ of college students. There are too many for us to measure all of them, but if it *were* possible, the distribution might look like the larger one in Figure 7-1, which is essentially the same as Figure 2-1 (page 7).

But a second distribution also is depicted in Figure 7-1. It is the sample that we have drawn from the population. It is somewhat out of scale, for we should probably never be able to measure that large a proportion of the total population; but the drawing will suit our present purpose very nicely. The point is, of course, that the sample mean $(\bar{X})$ is not always the same as the population mean (μ_X).

Now comes the fun part. So far, we have only one sample—and indeed that is precisely what we would have in a "real life" situation. But this is not real life; this is an attempt to understand the *structure* that underlies events that occur in the real world. To enhance that understanding, we shall have to imagine something that doesn't actually occur—namely, that after measuring one random sample, we

---

[1]See also pages 13 and 17. The present chapter and the ones that follow it are concerned with "sampling" or "inferential" statistics, as distinguished from the "descriptive statistics" of Chapters Two through Six.

[2]The basic ideas in this section were introduced earlier in our discussion of the stability of the mean (pages 15–17). There they were presented in a most rudimentary way; now we can afford to be more precise, because since then you have mastered the concept of the standard deviation, which, as you will see in a moment, is essential to an understanding of standard error.

"throw it back," so to speak, and take another—and another and another, *ad infinitum*. If we were to do that, and if we were to obtain a statistic—say a mean—from each sample before we threw it back, we should eventually acquire a very large number of means, all of them drawn from the same population.

As noted above, those means are not all alike. As a matter of fact, if we plot them in a frequency distribution, we find that the distribution is normal. It is a distribution of *means*; like any distribution, it, too, has a mean (the mean of the means). If we have, in our hypothetical investigation, taken an infinite number of samples, the mean of the means will be equal to the population mean.

So we have found that which we sought—pinned it down precisely! Unfortunately, however, it is only in fancy that we can take an infinite number of samples. In real life, we get only one; so how does all this really help? It helps because although a knowledge of the sample mean will never tell us what the population mean is, we *can estimate* from the variability (and the size; see pages 70-71) of the sample what the variability of a distribution of sample means would be. Our estimate of the standard deviation of a hypothetical distribution of sample means is called a *standard error* ($SE_{\bar{x}}$),[3] and it can be estimated if you know the size and variability of the sample

$$SE_{\bar{x}} = \frac{S}{\sqrt{N}}$$

Other Applications

Throughout the foregoing discussion of "standard errors," actually only one—the standard error of the mean—has been presented. That limitation was imposed because, although standard error may be a difficult concept to grasp initially, once it is understood in one situation it can readily be applied to others; the same logic applies to them—to the standard error of a standard deviation, of a correlation coefficient, of an obtained score, and many others. In every case, a *small* standard error tells us that the sample "statistic" is a reliable estimate of the corresponding population "parameter."

---

[3]Although it describes a hypothetical *population* rather than an actual sample, the standard error is regarded as a statistic rather than a parameter, because it is estimated from the characteristics of a sample.

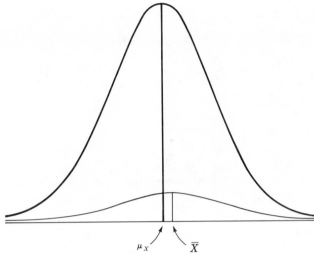

μ_x $\overline{X}$

FIGURE 7-1. *Distributions of IQ's of college students—population and sample.*

Confidence Levels

We now have a powerful tool to use in our attempt to locate the population mean, because now we can try some hypotheses and discover the probabilities of their being valid. Figure 7-2 indicates graphically how that might be done in the case of the mean IQ of college students if the standard error of the mean were $SE_{\overline{x}} = .102$.

Essence of the concept

The obtained mean IQ is 115.5. Figure 7-2 shows how that statistic plus the standard error of the mean can be used to answer the following question: "Within what limits may we be reasonably sure that the population mean resides?" The interval that is enclosed by those limits is called a *confidence interval,* and the "reasonably sure" in the above question is expressed quantitatively as a *level of confidence*.

You will recall that there is a variance in the means of random samples from the same population. That being the case, if the population mean were, say, .2 points *below* the obtained sample mean (Figure 7-2B), how many of a hundred sample means would be as *high* as the one we got? In the drawing, you can see that .025, or 25 out of every thousand, sample means would be that high. We can now state that

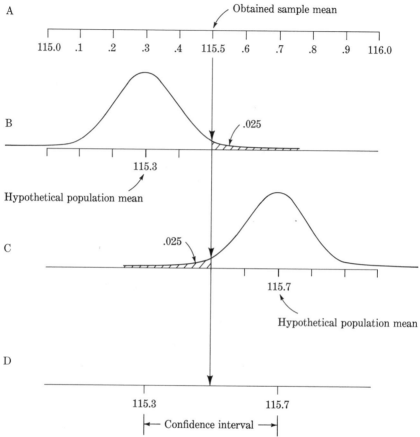

$$\text{SE}_{\bar{x}} = .102$$

A

Obtained sample mean

115.0 .1 .2 .3 .4 115.5 .6 .7 .8 .9 116.0

B

.025

115.3

Hypothetical population mean

C

.025

115.7

Hypothetical population mean

D

115.3 115.7

|←— Confidence interval —→|

FIGURE 7-2. *Establishing confidence interval at .95 level of confidence.*

the population mean is not lower than 115.3, and we can state it with a level of confidence of 1.000 minus .025, or .975.

The lower limit is only half of what we need to define a "confidence interval"; but with that behind us, the rest is easy. All we do is to repeat the above process, except this time we will test the hypothesis that the population mean is *above* the obtained mean. Now we ask, "If the population mean were .2 points above the obtained sample mean (Figure 7-2C), what is the probability that a sample mean as *small* as ours would occur?" Again the probability is .025.

The testing of our first hypotheses informed us that there is a probability of only .025 that the obtained mean is lower than 115.3; the second test yielded a probability of .025 that it is higher than 115.7. Now let us put the two hypotheses together: "What is the probability that the population mean lies outside the interval 115.3 − 115.7?" The answer is, of course, .025 + .025, or .05; so if we say that the population mean *is* somewhere within that "confidence interval," we can do so at a "confidence level" of 1.00 minus .05, or .95.

Some exercises to consolidate the concept

The deviation of an obtained mean from a population mean is a deviation score; $SE_{\bar{x}}$ is essentially a standard deviation. A deviation score divided by a standard deviation is a standard score. So we can divide our imagined deviation by our hypothetical standard deviation and obtain a standard score. In Figure 7-2B, the "deviation score" is .2 and the "standard deviation" is .102; the resulting standard score is $.2/.102 = 1.96$. If we were to enter a "normal curve table" with a standard score of 1.96, we would find that the "area in the smaller portion" of the curve is indeed .025. If you were learning to *do* statistics, there would be such a table at the back of this book, and you would get much practice in using it. However, since your objective is to *understand* statistics, it is better that you deal with such problems in graphical terms. The following exercises will give you several opportunities to do just that.

Trace the curve in Figure 7-2 on a piece of paper, take a pair of scissors, and cut around it so that you have a distribution that can be moved about in order to test various hypotheses. Try the limits of 115.4–115.6, 115.2–115.8, 115.1–115.9, and 115.0–116.0; estimate in each case the probability first that the obtained mean is *outside* the interval, then that it is inside. You can do that for any interval you wish; but in a real-life situation, you would decide in advance how much risk of error you are willing to accept and conversely what confidence level you would require. Then you would find the score limits that correspond to that level of confidence.

In other words, you would do just what I did in Figure 7-2. I decided initially that I wanted to illustrate a confidence level of .95, which leaves a probability of .05 that the obtained mean lies *outside* the confidence interval. Half of that .05 would occur above and half below that interval, so I proceeded as follows:

1. First, I slid the hypothetical distribution of means downward until the vertical line from the obtained mean cut off just .025 of the distribution. Then I designated the mean of that distribution as the lower limit of my confidence interval; if the population mean were any lower than that, less than .025 of repeated samples could be expected to have means as high as the one in my sample.

2. Having established the lower limit, I repeated the procedure in order to find the upper. I asked how high the distribution of sample means would have to be for just .025 of them to be below the obtained mean of my sample.
3. Having found that, I had both limits of the confidence interval that corresponds to a confidence level of .95.

Effect of N on Standard Error

We have already noted that the variability of a hypothetical distribution of means ($SE_{\bar{x}}$) is estimated from the variability of an actual distribution of individual cases (S). But there is another factor that is

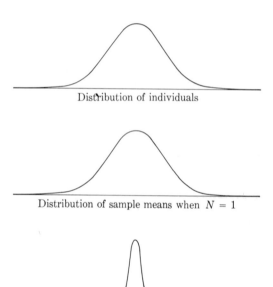

Distribution of individuals

Distribution of sample means when $N = 1$

Distribution of sample means when $N = 25$

Distribution of sample means when
N = The entire population

FIGURE 7-3. *Distributions of individuals and of the means of samples of three different sizes.*

extremely important in making that estimate: the *number* of individual cases (N) in the sample.

The relation of N to $SE_{\bar{x}}$ may be less comprehensible initially than that of S to $SE_{\bar{x}}$, since the latter relation is between two forms of the same concept; but the relation to N can be quickly illuminated by means of a few simple diagrams. The top drawing in Figure 7-3 represents the actual distribution of an entire population; each "case" in the distribution is an individual member of that population. The three drawings below it are hypothetical distributions; each case within each of them is the mean of a sample from the above population of individuals. The differences among the three distributions are striking— and those differences are a function of the differences in N.[4]

You may have been surprised to see that the first of the hypothetical distributions of means was exactly the same as the original distribution of individual subjects. But a moment's reflection will convince you that it could be no other way, because when $N = 1$, every sample *is* an individual subject. Similarly, you can see that if every sample included the entire population, the variability among sample means would have to be precisely zero. The middle drawing represents an intermediate situation in which every sample has an N of 25. In every case, the selection of samples is random, and each sample is returned to the population before the next one is taken.

Expectancy Tables

We have been describing ways in which sampling errors might be identified and quantified. We have done so in terms of the probability that our pronouncements about a population are correct. There are many occasions in contemporary social science research in which a relation between *two* variables is the focus of interest. On such occasions, various correlation techniques are applied; and a standard error can be estimated for a Pearson r, for example, as appropriately as it can for a mean.

The expectancy table as a scatterplot

If, however, there is a necessity of communicating "relational" information to a person who is untrained in statistics (for instance, a high school student or his parent) some other way must be found—

---

[4]The height of a drawing of a hypothetical distribution is arbitrary; because, in such a distribution, N is infinite.

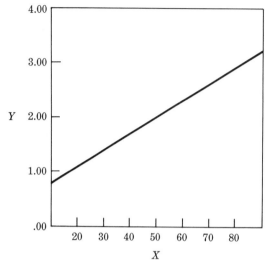

FIGURE 7-4. *Regression of Y on X when
r = .60.*

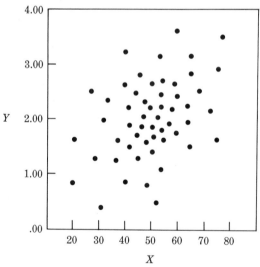

FIGURE 7-5. *Scatterplot of the same data.*

some display that is concrete enough to be grasped without previous
study. A "scatterplot" (see pages 37–42) is one such display. You will
recall that in seeking the Pearson r we were really attempting to define

TABLE 7-1 | *Expectancy of Frequencies*
(*Figure 7-5*)

Y	10–19	20–29	30–39	40–49	50–59	60–69	70–79	80–89
3.00–3.99				1	2	1	1	
2.00–2.99		1	2	7	7	3	2	
1.00–1.99		2	3	7	7	2	1	
.00– .99		1	1	2	1			

X

the slope of the regression line. That slope is related to the amount of scatter in the plot, but now we are going to *emphasize* scatter. Figure 7-4 shows the regression of Y on X when r = .60. If X were an aptitude test score and Y were grade-point average at Cerebral University, Figure 7-4 would tell a student with a score of 57 that if he goes to college his GPA will be 2.20.[5] Figure 7-5, on the other hand, tells him that although a 2.20 GPA may be the central tendency of students who score 57 on the test, it is not the only possibility.

An *expectancy table* functions in the same way but is easier to use than a scatterplot. The space in the table is divided on two dimensions into rows and columns, thus forming an array of cells. If the entry in each cell is a frequency, then the scatterplot is almost exactly duplicated, as in Table 7-1. Usually, however, the frequencies in each column are transformed into percentages of the column sum, as in Table 7-2.

Using an expectancy table

Now you can tell your student not only his most likely GPA in college, but also his chances of doing better (or worse) than the most probable performance. Joe Cawledge's score on Test X is 57; Table 7-2 says that of all students who previously scored in the 50's and subsequently have attended Cerebral University, 6 percent have earned a GPA below 1.00, 41 percent have been in the interval 1.00 through 1.99, another 41 percent between 2.00 and 2.99 and 12 percent achieved

[5]Grade points are not standard scores, of course. I have equated the X and Y scales arbitrarily for the purpose at hand.

TABLE 7-2 | *Expectancy of Percentages*
(Figure 7-5)

Y	10-19	20-29	30-39	40-49	50-59	60-69	70-79	80-89
3.00–3.99		0	0	6	12	17	25	
2.00–2.99		25	33	41	41	50	50	
1.00–1.99		50	50	41	41	33	25	
.00– .99		25	17	12	6	0	0	

X

a 3.00 or better. That kind of information would be comprehensible to the interested layman. Indeed, it may be helpful even to the experienced tester or counselor; for although he knows immediately that any *r* below a magnitude of 1.00 implies some scatter, it is often difficult for him to visualize the *amount* thereof. The expectancy table gets right down to cases.

A refinement that is often made in the interpretation of such a table is the derivation of *cumulative* percentages. Table 7-3, for example, implies not that Joe Cawledge has a 41 percent chance of achieving a GPA between 2.00 and 2.99, but that the probability of his doing *2.00 or better* is .41 + .12 = .53. That sum is often more useful than either probability by itself; in fact, that is so often the case that many expectancy tables are made up entirely of cumulative percentages. Table 7-3 is based upon the same data as 7-2; the only difference is that all the percentages are cumulative. That makes it less flexible than Table 7-2; but, for the many persons who are interested primarily in cumulative percentages, it is easier to use.

Summary

Every measure is derived from a sample, and although samples are, by definition, representative of the population from which they are drawn, they differ, one from the other, even when drawn from the same population. If there are no *systematic* deviations of the characteristics of the sample from those of the parent population, we say that the deviations are *random*, and we refer to them collectively as "error" variance.

Now if there is going to be error in our measurement, it behooves us to know its magnitude—or rather, its *probable* magnitude—for we might otherwise be more confident than we should of the statements that we make about the population from which any sample has been drawn. The most common way of quantifying error is to compute a *standard error* for whatever statistic we wish to cite.

A standard error is an estimate of the standard deviation of a hypothetical distribution of values that would be obtained for a given statistic if repeated samples were drawn from a single population. If the statistic in question were a mean, for example, we could estimate the variability of a distribution of means of successive samples of the same size from the same population. The standard deviation of that distribution is estimated by the *standard error of the mean.*

The standard error of the mean is important because with it we can mark off the *limits* within which the population mean probably lies, and we can ascertain what that probability is. In technical terms, we can state the *level of confidence* of our hypothesis that the population mean resides within the specified *confidence interval.*

The size of the standard error of the mean is a function of the variability of the sample, as indicated above. But it is also a function of the *number of individuals* in the sample. N varies all the way from *one* individual at one extreme to *all the individuals in the population* at the other. If each sample includes only one individual, the standard error is as large as the standard deviation of the population would be if we could get it. If the sample size is the same as that of the population, the standard error is zero, because repeated samples would all have the same mean. Other sample sizes have intermediate effects on the standard error.

TABLE 7-3 | *Expectancy of Cumulative Percentages (Figure 7-5)*

Y		10–19	20–29	30–39	40–49	50–59	60–69	70–79	80–89
3.00–		0	0	6	12	17	25		
2.00–		25	33	47	53	67	75		
1.00–		75	83	88	94	100	100		
.00–		100	100	100	100	100	100		

X

The standard error of the mean has been used here as an illustration of the error concept. Everything that has been said about it can be said also of the standard errors of other statistics.

Finally, there is another way to deal with "error" in situations that call for making predictions of an individual's position on one variable from a knowledge of his position on another. An *expectancy table* can inform a person who has a given score on the predictor variable (say, a scholastic aptitude test) what his chances are of achieving each of many possible scores on the criterion measure (say, grade-point average in college). The table thus deals explicitly with that which a correlation coefficient only implies—namely, the deviation of criterion measures from their regression line.

SIGNIFICANCE OF A DIFFERENCE BETWEEN TWO MEANS

Often, both in basic and applied research, it is important to know whether or not two populations are different from each other. From the point of view of the researcher, the question might better be stated, "Do the two samples that I have measured represent two populations, or are they merely two samples from the same population?" The true answer to that question is either "yes" or "no." He can never know the true answer, however; so his must be stated in probabilistic terms. The researcher's answer depends on two factors: (1) the size and direction of the obtained difference between the means and (2) the variability of a hypothetical distribution of differences that have been obtained in the same way that his was.

One way to illustrate the effect of variability (sampling error) on his answer is to analyze carefully a single example. In the following section, we shall examine an experimental study in the field of education, from the design of the experiment to the announcement of the results, concentrating throughout on the basic ideas rather than the computations.

An Example

Imagine that we have invented a new—and hopefully better—method of teaching French grammar to American high school students who have no knowledge of that language. Is it really better than the prevailing method? To find out, we take two samples from the population "Naive American High School Students," make sure that the two are initially random samples from a single population, and teach one group (hereafter to be known as the *control group*) by the traditional method and the other (the *experimental group*) by the new method. Then, after 150 hours of teaching time, we test both groups and compare their mean scores. The objective of our study is to ascertain whether the two are *still* "random samples of the same population" insofar as their knowledge of French grammar is concerned.

Imagine that the difference between the means is 10 points. Is the obtained difference significant? Is it large enough that we may reject the hypothesis that the control and experimental groups remain, after the teaching as they had been before, random samples from a single population with respect to knowledge of French grammar? That single population might be labelled "American High School Students Who Have Had 150 Hours of Instruction in French Grammar." Our hope is that, instead, there are *two* populations—one superior to the other with respect to knowledge of French grammar—and that the experimental group represents the superior population.[1]

Before we can be sure of that, however, we must disprove the hypothesis that the two groups are but two samples of a *single* population. That is the hypothesis of no difference, or the *null hypothesis*, and our test of significance is really a test of the null hypothesis: we attempt to disprove the hypothesis that there is no real difference between the groups—or rather, that the observed difference is merely a chance difference resulting from ordinary sampling error.

That may sound easy, but there is a complication. Like virtually every conclusion that emanates from statistical reasoning, this one must be stated not in absolute form, but in terms of probability. We'll not get a "yes" or a "no" out of our statistical test. Rather it will give us the probability that we are rejecting a true null hypothesis, and although that probability could conceivably approach zero, we'll

---

[1]This is admittedly very abstract, because you have to imagine here a population that does not at the moment exist—that is, a population made up of an infinite number of American high school students who have been taught French grammar by the new method.

never get a flat-out "no."[2]

What we do, then, is to adopt arbitrarily some level of probability that is *an acceptable approximation* of zero, and when a test reveals a probability below that level, we reject the null hypothesis. (After that, we may entertain others.) Whether or not a particular difference—like the 10 points we obtained between our control and experimental groups after the language instruction—meets that preset criterion depends upon *the variability of a hypothetical distribution* in much the same way that the size of the "confidence interval" did when we were discussing the stability of the mean. It might be worth your while to review that section before proceeding with this one; it begins on page 67.

In fact, a knowledge of the standard errors of the means of both groups is essential to our present purpose. This time, however, we are dealing primarily with a hypothetical distribution of *differences between* means—in this case, between the means of groups of American high school students taught by two different methods.

Test of Significance: The z Ratio

Is a difference of 10 points sufficiently large to be significant, or is it small enough that it might easily have occurred by chance—by ordinary sampling error? The answer is a function not only of the size of the difference itself but, very importantly, of the sampling errors—as indicated by the standard errors of the means—of the two groups being tested. The next six paragraphs, together with Figures 8-1 through 8-4, concern a situation in which the standard error of the mean is rather small and is the same for both groups.

Look at Figure 8-1. It shows two hypothetical distributions of sample means on a scale of French grammar scores; the solid line represents control groups, the dotted line "experimentals." They occupy the same space because the drawing was made on the hypothesis that the two kinds of groups are in reality *not* "two kinds" with respect to achievement in French grammar, but are merely two sets of random samples from a *single* population. That is, of course, the "null hypothesis," and we shall try our best to disprove it (or rather, to make it untenable).

---

[2]If we do *not* reject the null hypothesis, we must regard it as tenable (our data offer insufficient evidence against it) but not necessarily as true. Other hypotheses are also tenable in this sense (for instance, the true difference *could* be identical to the one we obtained from our samples).

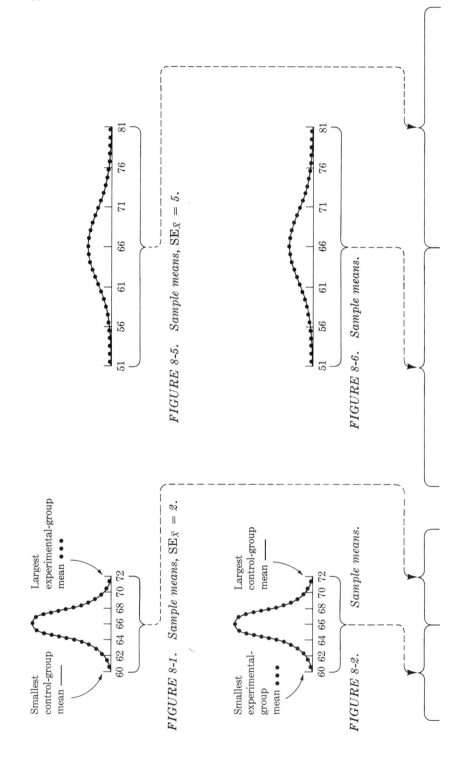

FIGURE 8-5. *Sample means,* $SE_{\bar{x}} = 5.$

FIGURE 8-6. *Sample means.*

Largest
experimental-group
mean ●●●

60 62 64 66 68 70 72

Smallest
control-group
mean —

FIGURE 8-1. *Sample means,* $SE_{\bar{x}} = 2.$

Largest
control-group
mean —

60 62 64 66 68 70 72

Smallest
experimental-
group
mean ●●●

FIGURE 8-2. *Sample means.*

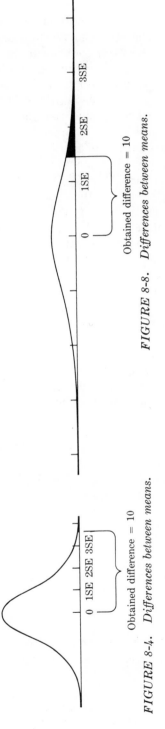

FIGURE 8-3. *Differences between means,*
$SE_{\bar{X}_e - \bar{X}_c} = 2.83.$

Obtained difference = 10

FIGURE 8-4. *Differences between means.*

NOTE: *Figures 8-1 through 8-4 show test of
significance of differences when* $SE_{\bar{X}_c}$ *and* $SE_{\bar{X}_e}$
are small. Distributions are all hypothetical.

FIGURE 8-7. *Differences between means,* $SE_{\bar{X}_e - \bar{X}_c} = 7.07.$

Obtained difference = 10

FIGURE 8-8. *Differences between means.*

NOTE: *Figures 8-5 through 8-8 show test of
significance of differences when* $SE_{\bar{X}_c}$ *and* $SE_{\bar{X}_e}$
are large. Distributions are all hypothetical.

In order to test the null hypothesis, it is necessary to think in terms of a scale not of scores but of *differences between* scores—specifically, of differences between control- and experimental-group means. Look again at Figure 8-1 and imagine taking pairs (one control and one experimental group in each pair) of means at random from the two distributions. Most of the intra-pair differences would be close to zero; but a few—just by chance, remember—would be substantially larger. In fact, if you were to compare the largest experimental-group mean with the smallest of the "controls," the difference would be 6 standard errors, or 12 raw-score points.

That difference, $\bar{X}_e - \bar{X}_c$, is in the direction we have predicted, so we'll call it "positive." But if all the samples are drawn from the same population, as the null hypothesis would have it, there should be as many "negative" differences as there are positive, and the largest of them should be just as impressive as the positive difference we just found. Figure 8-2 represents the same two distributions as Figure 8-1; if you will compare the *lowest* experimental with the *highest* control group, you will indeed discover that the difference is "-12."

Figures 8-1 and 8-2 are the same distributions analyzed in two ways: one as positive, the other as negative differences. Figure 8-3 combines "positive" and "negative" differences into a distribution of differences. Positives are on the right, negatives are on the left, and the mean is zero—the null hypothesis again. If we were to compute an infinite number of differences between pairs of means, selecting each pair at random from a single population, that is about what the distribution would look like.

With that much behind us, the testing of the null hypothesis should be relatively easy. All we have to do is place our obtained difference within the hypothetical distribution of differences, as in Figure 8-4, and we can see immediately that only a very few positive differences[3] that large occur simply by chance among samples taken from a single population.

By inspection it appears that hardly any of such differences would be as large as the one we obtained in our experiment; we are justified, therefore, in rejecting the null hypothesis as untenable. Formula 8-1 offers a more precise estimate of the variability of differences between means.*

$$\text{SE}_{\bar{X}_e - \bar{X}_c} = \sqrt{\text{SE}_{\bar{X}_c}^2 + \text{SE}_{\bar{X}_e}^2}$$

(8-1)

where $\text{SE}_{\bar{X}e - \bar{X}c}$ is the standard error of a difference between means, $S_{\bar{X}c}$ is the standard error of the mean for the control group, and $S_{\bar{X}e}$

---

[3]Asking only about *positive* differences constitutes a "one-tail test." A discussion of one- vs. two-tail tests begins on page 85.

is the standard error of the mean for the experimental group. Since our interest is in the underlying logic of the significance test rather than in the precision afforded by the formula, we shall pass by the latter with only the comment that, like Figures 8-1 through 8-4, the formula shows that the variability of random *differences between* means taken in pairs $\mathrm{SE}_{\bar{X}_e - \bar{X}_c}$ is proportional to variability *among* means taken singly $S_{\bar{X}_c}$ and $S_{\bar{X}_e}$.

To illustrate that point further, and to emphasize the importance of $\mathrm{SE}_{\bar{X}}$ in determining the fate of the null hypothesis, turn back to page 80 and see how our 10-point obtained difference would have fared had the sample means been less stable (Figures 8-5 through 8-8). Whereas in Figure 8-4 our difference was 3.53 standard errors ($z = 3.53$) above the mean and had a probability of .0002, in Figure 8-8, a difference of ten is only 1.41 standard errors away from the mean and has a probability of .0793. In the latter case, if we were to take 1000 pairs of samples at random from *a single* population, 79 of the intra-pair differences would be at least as large as ours. Had we obtained our difference in those circumstances, we'd have had to accept the null hypothesis as tenable. Incidentally, do not be disturbed if you cannot estimate the above values with this kind of precision simply by looking at the graphs; I did it by formula.*

All of the examples in this section have been treated graphically in order to expose the structure of their operations. Normally, however, those operations are done algebraically for greater accuracy. In the language-teaching study, the formula would be

$$z = \frac{(\bar{X}_e - \bar{X}_c) - 0}{\mathrm{SE}_{\bar{X}_e - \bar{X}_c}} \tag{8-2}$$

where z is the ratio of an *obtained difference* to the *standard error of a difference* between means in a distribution of such differences when the mean of the distribution is zero. In Figure 8-4, that ratio is 10/2.83; and in Figure 8-8, it is 10/7.07. The zero in the formula has no effect on the outcome; it is there only to remind you of precisely what it is that the formula is supposed to do; namely, to ascertain how far our obtained difference is from the mean of a distribution of differences, all of which are the results of sampling errors. The mean of such a distribution would indeed be zero.

Test of Significance: The t Ratio

We may regard Formula 8-2 as the basic formula for the test of significance because it clearly reveals the relation between the difference and the standard error of the difference. When that difference is obtained

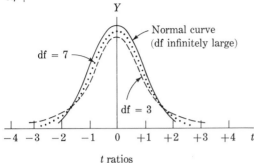

FIGURE 8-9. *Distribution of t ratios at three different sample sizes.* (*Adapted from* Introduction to Probability and Statistics, *5th edition, by Henry L. Alder and Edward B. Roessler. W. H. Freeman and Company. Copyright* © *1972.*)

from a small sample, however, it is not the *z* ratio that is tested for significance, but a ratio called "*t*." The *t* ratio is essentially the same as *z*, but it is referred for significance testing to a distribution that changes its shape (because the probability of obtaining the various scores in it changes) as *N* gets smaller. When *N* is larger than 30, the difference between *z* and *t* distributions is negligible; when *N* is infinitely large, the two are identical.[4] In Figure 8-9, it is clear that when the *t* distribution is used, the probabilities to be inferred from various placements on the base line are in many instances quite different if *N* is small than if it is large; most notably, in the former case extreme ratios make up a larger proportion of the distribution.

Because it is appropriate for use with either large *or* small distributions, the *t* test is used almost universally in place of *z*. But because *z* is a logical extension of concepts developed earlier, it is best to think of *t* as a modified *z*.

Significance Levels

We have seen above that in certain circumstances a difference of 10 points between two sample means might occur frequently by chance. However, our difference was obtained not in the circumstances depicted

---

[4]You will notice that in Figure 8-9, sample size is indexed by "df" (degrees of freedom) instead of by *N*. That is really all you need to know in order to follow this discussion. However, the concept of "degrees of freedom" is discussed briefly in a note in the back of the book (the note referring to Chapter Four, page 25).

in Figures 8-5 through 8-8, but in those of Figures 8-1 through 8-4; there it is clear that the probability is extremely small that, with respect to their knowledge of French grammar, our control and experimental groups are random samples of a single population. Now we may annouce, to an eager educational public, that our new method of teaching French grammar is almost certainly better than the traditional one—at least under the circumstances prevailing in our experiment. We proclaim that superiority at a *significance level* of .0002, because there is only that much of a probability that our proclamation is wrong —that we are rejecting a true null hypothesis.

It may seem to you that the "level of significance" in this instance should be 1.000 − .0002, or .9998. If it does, I agree with you; but tradition holds that the level must be expressed as *the probability that a true null hypothesis is being rejected*. That means that the *lower* the significance level, the *higher* is our confidence that the effect we have observed is real.*

Tradition also provides us with two levels that are recognized as significant and very significant, respectively. A "significant" difference is less than .05, which is often written "$p \leq .05$." A "very significant" difference is one for which the probability of having occurred by sampling error is less than .01; it is sometimes written "$p \leq .01$." Finally, in tables and other places where the briefest possible abbreviations are used, a "significant" difference sometimes is designated simply by the letter "S" and a "very significant" one by "VS." In practice, those two levels are often used but also often ignored. For example, a researcher who had obtained as impressive a difference as that between our control and experimental groups in the language-testing study would be unlikely to hide his light under the bushel "less than .01"!

One- vs. Two-tail Tests

There is one last thing that should be said about the testing of significance and the levels of confidence that emanate therefrom. In Footnote 3 (page 82), I told you that what we were doing there constituted a *one-tail test* of the significance of our obtained difference of 10 points; we were saying that there was only a .005 probability that a difference that large *and in the expected direction* would occur by chance. The alternative is a *two-tail test*, which is appropriate whenever *we have not predicted the direction* of the difference. If you will look back at Figure 8-3 for a moment, you will see that the probability of a difference of 10 in *either* direction is *double* that of a difference in *one* direction. That means that a particular difference conceivably could fail the significance test if there has been no directional prediction and pass it if a prediction has been made (provided, of course, that the outcome matches the prediction). But we are not free to select either test arbitrarily; if we have no reason to expect a difference in one direction rather than another, we are obligated to use the two-tail test. It is only when we *do* have such a reason, *when we can make a prediction in advance,* and when the results confirm the prediction, that we are justified in using the one-tail test.*

Practical vs. Statistical Significance

The length of this section barely exceeds that of a long footnote, but it contains an important notion that might easily be overlooked if it were not emphasized. The point is that even though a difference is shown to be statistically significant, it may not have any *practical* significance.

The result of our language-teaching experiment was very convincing; there can be almost *no* doubt that the difference between the two groups at the end of the teaching period was real. But how does that information affect the decisions of a school administrator who must decide whether to adopt the new method? That depends on many considerations, some of which have nothing to do with statistics. It depends to a great extent on how costly the new teaching method is to implement—whether it requires expensive equipment or specially trained teachers. The administrator's decision may also depend on the absolute size of the difference. That may seem a contradiction of everything we've been saying up until now; and it is, unless we keep in mind the distinction between statistical and practical significance.

The joker is in that standard-error term—more specifically, the importance of N in determining its size. If you analyze either Figures 8-1 to 8-8 or Formula 8-1, you find that the size of the standard error of the difference is proportional to the standard errors of the means of the two samples. And the size of a standard error of the mean is determined partly by the variability in the sample and partly by its *size*

$$\text{SE}_{\bar{x}} = \frac{S}{\sqrt{N}}$$

Now, the significance of an obtained difference depends upon the ratio between that difference and the standard error of the difference

$$\frac{\bar{X}_1 - \bar{X}_2}{\text{SE}_{\bar{X}_1 - \bar{X}_2}}$$

That means that if our samples are large enough, *any* difference will be statistically significant, no matter how small it may be; and while a *z* ratio of 3.53 indicates that the difference, no matter how small, is probably genuine, a school administrator might decide that a very *small* difference would not be worth its additional cost, even if it were *absolutely* reliable.

Summary

There are times when it is important to know whether two groups are different from each other—or rather, whether they *represent populations* that are different from each other. In this chapter, we have compared the performance of two groups of students after exposure to two different teaching methods. One group, taught by the traditional method, was called the *control group*; the other, taught by a new method, was the *experimental group*. The experimental group outscored the control.

But two groups given the *same* treatment could have scored differently by *chance*, and in such a case we should have to admit that the obtained difference was a *sampling error*—that the two groups are really only two samples from the same population with respect to the performance being tested. We must consider the posibility that our control and experimental groups are like that, too—that the difference we obtained was due entirely to sampling error and that the two groups are really only two samples from a single population. That possibility is known as the hypothesis of no difference, or *null hypothesis*.

The general question, then, is really, "How large does an obtained difference have to be before we are justified in rejecting the null hypothesis?" The procedure by which that question is answered is known as a *test of significance*. In the case at hand, we ask a more specific form of the question, namely, "Is our obtained difference large enough to justify a rejection of the null hypothesis?"

The answer will not be a simple "yes" or "no"; it must be given in terms of *probability*. What is the *probability* that a difference as large as ours would occur if there were *no* real difference in the effectiveness of the two teaching methods? Acceptable levels of probability are somewhat arbitrary, but two such *levels of confidence* have traditionally been set at .05 and .01. However, a researcher who obtains a null hypothesis probability much lower than .01 will probably report it exactly; because the lower it is, the more impressive is the outcome of his experiment.

Is it possible to ask all of the above questions about any measurement of any two groups of subjects? We could use the test of significance simply to explore—to search out differences worthy of further investigation. But in the case of the two teaching methods described above, we were not exploring; we *expected* the experimental group to be better than the control. That is an important distinction, because if we can state our expectancy in advance, we are privileged to use a *one-tail test* instead of a *two-tail test* of significance. If we are exploring, we must ask, "What is the probability of a sampling error this large *in either direction?*"; whereas, if we have stated our expectancy, we may ask instead, "What is the probability of obtaining this large an error *in favor of the experimental group?*" Since the probability of a difference in *one specified* direction is just half that of the same amount of difference in *either* direction, the one-tail test is twice as sensitive as the one we would be obliged to use if we were merely exploring; a difference half as large will qualify at whatever level of confidence we have prescribed.

NINE

MORE ON THE TESTING OF HYPOTHESES

A chapter with this title could easily occupy as much space as all the rest of the book put together. It will not, because I have selected just two tests of significance as illustrations. The others will be mentioned only briefly, if at all. For our purposes, it is not necessary to describe those others. In fact, probably the most important contribution that I can make to your understanding of all of them can be stated without *any* illustrations. The fundamental idea is this: each of them is conceived as *the testing of a null hypothesis*—a hypothesis of no difference.*

The difference that emerged from our experiment with two methods of teaching French grammar was a difference in *amount*. We asked whether an innovative method produced students who were more able than those produced by the more traditional method with which it was being compared. More technically, we asked whether the difference we obtained was attributable to the difference in teaching methods or whether it was merely the result of sampling errors—that is, we tested the null hypothesis. But we could have asked a different question: Given some clear criterion of "passing," does the innovative method produce fewer failures than the traditional? That question is posed in terms of *frequencies*, and although the test described in Chapter Eight can be adapted to deal with frequencies, proportions, and probabilities, there is another one that is more often used in such cases. So the first part of the present chapter will examine a test of the null hypothesis in which all of the data are in the form of frequencies.

There is another feature of the example that we used in Chapter Eight that is not characteristic of all experiments: only two "treatments" were compared. What should we do if we wanted to assess the relative effectiveness of *several* teaching methods? Or what if we suspected that one treatment method might be effective in the hands of one teacher and another might be superior when used by a different teacher? The second section of the chapter will describe a kind of analysis by which both of those questions may be answered.

Notice that every "test of significance" is a test of the null hypothesis. Conversely, any test of the null hypothesis is a test of significance; and since any observed difference could conceivably have resulted from sampling errors, any difference may be put to the test of significance.*

Comparison of Frequencies: Chi Square

In the introduction to this chapter, I mentioned that the scale of scores that had been used earlier in our investigation of teaching methods could have been reduced to two scores only: passing and failing. I pointed out that under those circumstances, the data would probably be in the form of frequencies—that is, how many subjects passed in each of the two conditions.

In practice, however, test-score scales are seldom reduced to two class intervals; to do so would be to throw away information. Significance tests designed to deal with frequencies, proportions or probabilities are usually applied in situations that do not produce finely divided scales in the first place. In a public opinion poll, for example, respondents are usually asked a question to which they are expected to answer either "yes" or "no," "for" or "against," and so forth.[1]

Let us consider briefly the reasoning behind a significance test of a set of data from a public opinion poll. Imagine that a particular male candidate for public office has considerably more sex appeal than

---

[1]There are ways of obtaining more finely graded responses, but they are more cumbersome to administer than the single question. Furthermore, a relatively crude index is often the most appropriate to the circumstances. For example, in predicting the outcome of an election, every mark that a voter makes on his ballot represents what is essentially a "yes-no" decision.

TABLE 9-1 | *Arrangement of Data in Election Problem*

VOTE FOR MR. SEX

	Yes	No
SEX OF VOTER M		
F		

another but seems to differ very little from his only opponent (another male) on any other dimension. He wins, and we wonder whether his sex appeal has played a part in his victory.

If we assume that the voters of both sexes have a basically heterosexual orientation, we may get an informative answer by rephrasing the question: "Did women vote for Mr. Sex in significantly larger numbers than did men?" The data can be arranged in a 2 × 2 table (see Table 9-1). If there are in this election 20,000 voters—10,000 male and 10,000 female—of which we have a one percent representative sample, if Mr. Sex has received 60 percent of all the votes cast, and if there are *no* differences between men and women with respect to voting behavior in this election, the table will look like Table 9-2.

You should recognize that last "if" as the null hypothesis; our test of significance will be an attempt to show that it is untenable. To do so, we shall have to demonstrate (1) that our *obtained* frequencies differ from those that would be *expected* if there were no real difference between the voting behaviors of men and women and (2) that the difference is larger than we should be likely to get through sampling errors.

TABLE 9-2 | *Expected Frequencies (Null Hypothesis)*

VOTES FOR MR. SEX

SEX OF VOTERS		Yes	No	Total
	M	60	40	100
	F	60	40	100
	Total	120	80	200

TABLE 9-3 | *Obtained Frequencies (From Actual Sample)*

		VOTES FOR MR. SEX		
		Yes	No	Total
SEX OF VOTERS	M	50	50	100
	F	70	30	100
	Total	120	80	200

The division of votes in our sample turns out to be as in Table 9-3.

Now, in Table 9-4, we'll put expected and obtained frequencies together, the better to compare them. It is clear that our candidate received a higher proportion of the female vote than of the male. But is the difference *significant*? How likely is it that a difference (actually a set of four differences in Table 9-4) as large as that would occur in a random sample if there were no difference at all within the population of voters.

To find out, we may use a statistic called Chi Square (χ^2). The general formula is

$$\chi^2 = \Sigma \frac{(f_o - f_e)^2}{f_e} \qquad (9\text{-}1)$$

where f_o = obtained frequency and f_e = expected frequency. Focus on one part of that formula for a moment

$$\frac{(f_o - f_e)}{f_e} \qquad (9\text{-}2)$$

TABLE 9-4 | *Obtained Minus Expected Frequencies*

		VOTES FOR MR. SEX	
		Yes	No
SEX OF VOTERS	M	50 −60 −10	50 −40 10
	F	70 −60 10	30 −40 −10

and you will grasp intuitively the fundamental nature of Chi Square. The greater the deviation from the expected frequency (expressed as a proportion of that frequency) the larger is Chi Square; and since the expected frequency is "expected" on the hypothesis of no difference, Chi Square is an index of deviation from that hypothesis. Tables are available in which one can find the significance level of any given Chi Square. In the present example, Chi Square is 8.34, which is significant at the .01 level; the probability that our obtained difference was due to sampling errors is less than one percent.*

Multimean Comparisons: Analysis of Variance

In Chapter Eight, we applied a significance test to a difference between two means. That is frequently what is needed in behavioral research, but it certainly is not always appropriate. Often an experimental design will call for more than two categories of the independent variable, and sometimes systematic changes are made in more than one independent variable.[2] Tests based on the z and t ratios described in Chapter Eight cannot be used to assess significance in either of those situations. But there is a technique that is appropriate to either or both of them; it is called *analysis of variance* (ANOVA).

One-way analysis of variance

In this section, we shall be concerned only with the former of the two designs mentioned above—namely, the one in which (1) there is one independent variable, but (2) there are more than two categories of that variable. Our example will be a simple extension of the experiment we used to illustrate the application of the z (or t) ratio. There, we had two groups, which we called "control" and "experimental"; here, we shall have six groups, and we shall refer to them as simply I, II, III, IV, V, and VI. There, we were comparing a new method of teaching French grammar with a traditional method; here, there are six different methods, one or more of which may be traditional. Let us say that the control and experimental groups of the earlier experiment are Groups I and IV of this one.

---

[2]An *independent variable* is one that is manipulated by the experimenter. (Mathematically, it is independent; experimentally, it is really a *manipulated* variable. But the mathematical term is always used, whether the context is mathematical or experimental.) A *dependent variable* is one whose values are determined (in cases in which the null hypothesis is properly rejected) by those of the independent variable(s).

We shall be wanting to know which of the six methods is (are) the most effective. We could proceed by comparing all possible combinations of groups, but that would be tedious, since it would require, in this problem,[3] 15 such tests to check all the possibilities. What we need is some kind of "survey" test that will tell us whether there is any significant difference *anywhere* in an array of categories. If it should tell us "no," there would be no point in searching further.

Such a test does exist. It is called the *F test* or *F ratio*. *F* is a ratio of two variances. If you will look back to Formula 4-3, page 25, you will find the standard deviation defined as

$$S = \sqrt{\frac{\Sigma x^2}{N}} \qquad (9\text{-}3)$$

The *variance* of a distribution, like its standard deviation, is a measure of variability. In the standard deviation formula, it is the part under the radical:

$$S^2 = \frac{\Sigma x^2}{N} \qquad (9\text{-}4)$$

It is the mean of the squared individual deviations, or, to put it another way, it is the square of the standard deviation.*

One important application of the *F* ratio is to the overall test of significance in analysis of variance. As always in a significance test, the null hypothesis states that all samples are random samples from a single population; and as always, we shall attempt to disprove that statement. Again, as in the *z* and *t* tests, it is a ratio that is being evaluated. But this time the critical ratio is not of a difference to its standard error; this time it is a ratio of two estimates of the population variance—one estimate from the means of the categories being studied, the other from individual scores *within* categories.

$$F = \frac{S_b^2}{S_w^2} \qquad (9\text{-}5)$$

---

[3]There are other reasons for using an overall test of significance—reasons that are more important in the long run than the saving of labor. First, any statistic based on *all* the data will be more stable (see Effect of *N* on Standard Errors, pages 70–71) than one which is based on only part of it, as would be the case if only two of the six methods were compared. And second, there are so many comparisons that some will be "significant" by chance. If there were a hundred such comparisons, five should show significance at the .05 level and one at the .01 level, even if there were no real differences at all.

where S_b^2 = variance "between" (among)[4] groups and S_w^2 = variance within[5] groups.

You will remember that back in Chapter Seven (especially pages 65–66) we used obtained individual scores to make an estimate of the variability of means in a population. Well, it can work the other way, too; we can use obtained means to make an estimate concerning the variability of individual scores in a population. The numerator in the *F* ratio is just such an estimate—*the population variance estimated from the variability of means of the categories* being studied;[6] the denominator is the population variance estimated from the variability of *individual scores within* those categories.[7] So we have in this ratio two estimates of the same population variance. The null hypothesis states that one of those estimates is as large as the other—that the ratio is 1.00.

If we are to demonstrate that some real difference does exist among the effects of our six teaching methods, we shall have to show that the ratio is *not* 1.00. Specifically, we shall have to show that the variance estimated from group means is greater than the variance estimated from individual scores within the groups. Understanding why that is so requires an analysis of the logic behind the *F* test.

Every estimate we make of population variance is an estimate of the variability of individual scores around a true mean. The closest thing we have to a true mean here is the *grand mean*. Every individual deviation from the grand mean can be thought of as two components: (1) the deviation of the individual score from its group mean and (2)

---

[4]The term "between-groups variance" is used no matter how many groups are being compared. (Statisticians are not punctilious grammarians.)

[5]This estimate is sometimes referred to as "residual," or even "error," variance (S_e^2). The latter term may be somewhat confusing, because most of the variance probably stems from factors that in different circumstances might prove to be effective independent variables. They are regarded as "error" in the context of a given investigation because they are random with respect to the independent variables in *that* experiment. They are the so-called "controlled" variables. There are two general strategies for "controlling" a variable in an experiment: (1) to hold it constant over all conditions of the independent variable(s) and (2) to randomize its variance. Both strategies are designed to achieve the same objective: prevention of any systematic relationships between the "controlled" variable and any independent variable. There usually are several controlled variables in an experiment.

[6]*F* is a statistic. You may wish to review pages 16–17 and 65–66, on the relation between a sample statistic and a population parameter.

[7]The terms "category" and "group" will be used interchangeably, with the understanding that the referent in each case is a set of values derived ultimately from the behavior of individual subjects collected into categories for the purpose of the experiment.

I   III   II   Grand mean   V VI   IV

FIGURE 9-1.   *Superimposed distributions of Groups I through VI.*

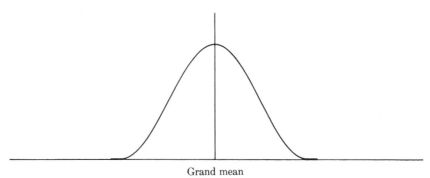

Grand mean

FIGURE 9-2.   *Superimposed distributions of Groups I through VI with differences between means eliminated.*

the deviation of that group mean from the grand mean. Now, if we could somehow eliminate the means component, we should have an estimate of *what the population variance would have been if there had been no differences* among the group means. That estimate *can* be made (by measuring the deviation of each individual score from its own group mean rather than from the grand mean); the result is the denominator of the F ratio.

If you will compare Figures 9-1 and 9-2, you will see immediately that the total variability of the six groups is smaller when the variability of their means is eliminated. (You see only *one* mean in Figure 9-2, because all six are in the same place.) That is why the denominator of the F ratio is smaller than the numerator if there are any differences at all among the group means.

Those group means do vary, almost always; the question is whether they vary enough to be significant in the sense that was developed in Chapter Eight. To find out, we compute a ratio. To provide the numerator of that ratio, we discover how much the *group means deviate from the grand mean.* For the denominator, we find out how much *individuals* deviate from their own *group* means.

The null hypothesis is that there is no variance among the means: hence it predicts that a population variance estimate that *includes* that variance (the numerator of the F ratio) will be the same as one which *excludes* it (the denominator); if there are no differences among the means, F will equal 1.00. Ratios larger than 1.00 can be evaluated by means of tables that give the probability that any given F would occur entirely by chance. That of course, is the *significance level* of the ratio, and it is interpreted in the same way as one which derives from a z or a t. We reject the null hypothesis if the probability of its occurrence by chance is sufficiently low; usually $p \leq .05$ will suffice.

After the F test

When an F test turns out to be significant, we know (with some specified degree of confidence) that there is a real difference somewhere among our means. But we don't know *where* it is.

The most obvious approach to this problem is to perform a t test on each difference, beginning with the largest and continuing until one test fails to achieve significance. That is not acceptable, however, for some of the same reasons that prevented us from using t in the first place. Take another look at footnote 3, page 94; the second reason cited there is the most cogent one for not using t after F. ("There are so many comparisons that some will be 'significant' by chance.") Some statisticians will approve the use of t in multivariate designs, either before or after analysis of variance, but only if the particular comparisons to be made are selected on a rational basis *before the data are collected*. It is the same requirement that was mentioned earlier in relation to the use of a one-tail test of significance (page 85); the reason for it is essentially the same, and the same controversy obtains.

Other tests have been devised for use in the post-F situation. All are attempts to disprove the null hypothesis, and all have been made more difficult to "pass" than t in order to compensate for the number of comparisons (and the concomitant increase in the probability that some will show significant differences by chance). Some of these tests are included in Appendix I under the title, Multimean Comparison Tests.

More complex designs

We have examined the logic of the F test using a "one-way" analysis of variance as an example. Others are possible: two-way, three-way, four-way, and so forth. My first impulse was to illustrate a two-way analysis by expanding our one-way example into a methods-by teachers design; each method would have been used by five teachers, making six methods $\times$ five teachers $=$ 30 treatment conditions. But that

TABLE 9-5 | *A 2 × 2 Factorial Design with Index Numbers of Subgroups of Subjects as Cell Entries*

RECENT EXPERIENCE

Failure Success

		Failure	Success	
STRESS	*High*	3	4	(Group "3-4")
	Low	1	2	(Group "1-2")

(Group (Group
"1-3") "2-4")

would have gotten us into problems that are inappropriate to a discussion of this kind. For example, the performance of a teacher using any one method probably would be affected by whatever experience he has had with the other methods; that would have taken us into the problem of counterbalancing the design. Also, the larger number of cells in such a matrix would have made your comprehension of *interaction* effects more difficult than necessary. So let's leave that one, with the passing comment that once the design problems have been solved and the treatments applied, data from such an experiment can be evaluated by analysis of variance techniques.

What we want now is the simplest design that could possibly be used to illustrate the basic principles of two-way analysis of variance—a "2 × 2 factorial" design. Also, we'll select independent variables that do not require a counterbalanced design. As in the preceding illustration, there will be no computations.

The dependent variable in this experiment is "persistence in the face of failure"—specifically, the amount of time spent on an insoluble problem. The two independent variables are (1) high vs. low stress in the subjects and (2) success vs. failure experience immediately before the testing period. The design is shown in Table 9-5 as a 2 × 2 matrix in which there are four cells, each of which represents one treatment condition—(1) low stress with failure experience, (2) low stress with success experience, (3) high stress with failure experience, and (4) high stress with success experience. (Find the corresponding index numbers in Table 9-5.) There are 25 subjects in each treatment condition; each subgroup is a random sample of a population of 10-year-old American males, and the question to be answered is whether they will *remain* random samples of a single population at the *end* of the experiment. The null hypothesis says they will.

TABLE 9-6 | Group Means (Minutes at Task) in
Two-Way Analysis. Outcome 1

RECENT EXPERIENCE

		Failure	Success	Total
STRESS	High	3 / 20	4 / 30	50
	Low	1 / 10	2 / 20	30
	Total	30	50	80

Several hours before the experiment begins, all subjects receive a painful electric shock "accidentally" while playing with some laboratory equipment. (As will become apparent later, it is important that they know what it is like to be shocked.) Then, just a few minutes before the persistence task is introduced, they all take a short paper-and-pencil test. The test is also the same for all subjects; however, half of them (Group 2-4) are told that their performances are successful and the other half (Group 1-3) that *their* performances are not.

Immediately after that experience, they are all introduced to the persistence task. The problem is insoluble, but the subjects don't know that. They are all told that they should do the best they can, but that they may leave at any time they wish. Then half of them (Group 3-4) are told that if they fail the test they will receive several shocks of the kind they had experienced earlier, while the other half (Group 1-2) are told nothing of the shocks; thus, we are manipulating stress as a second independent variable.

Now there are three advantages that analysis of variance offers over the z or t type of significance test: (1) it can compare the effects of more than two categories of an independent variable,* (2) it can compare the simultaneous but separate effects of two or more variables and (3) it can assess the *interaction effects* of two or more variables. The first of those advantages was illustrated in the preceding section (One-Way Analysis of Variance). The second can be attained by merely repeating the F test for the second independent variable. In the present case, we would do one F test for the "stress" *main effect* (Group 1-2 vs. Group 3-4) and another for "recent experience" (Group 1-3 vs. Group 2-4). The most interesting of the three features of ANOVA, but also the most difficult to understand, is the last-named: the ability of this kind of analysis to identify interaction effects.*

TABLE 9-7 | *Group Means (Minutes at Task) in Two-Way Factor Analysis. Outcome 2*

RECENT EXPERIENCE

		Failure	Success	Total
	High	3 / 10	4 / 30	40
STRESS	Low	1 / 20	2 / 20	40
	Total	30	50	80

The best way to explain interaction is to cite an example, so let's return to our experiment. Table 9-6 is like Table 9-5, except that the index number of each subgroup is in the upper left-hand corner of that subgroup's cell, and the number in the middle of the cell is the subgroup's mean score (time spent before quitting). The data in the table show what look like two substantial main effects. (Whether or not they are *significant* depends, of course, on the amount of variance within the two groups being compared in each case; let us assume a sufficiently small amount.) Group 3-4 (50 minutes) is more persistent than Group 1-2 (30 minutes), and Group 2-4 (50 minutes) is more persistent than Group 1-3 (30 minutes). High stress produces more persistence than low, and success more than failure. But there is no interaction.

By contrast, imagine that the outcome of the experiment were that depicted in Table 9-7. There is *one* main effect (success vs. failure), and *there is interaction* because increasing stress (from low to high) has a different effect on subjects in the "failure condition" (Group 1-3) than it does on those in the "success condition" (Group 2-4). Similarly, success (as opposed to failure) has a different effect in the "high-stress condition" (Group 3-4) than it does in the "low-stress condition" (Group 1-2). The interaction effect may be seen more clearly in the graphical representation of the two outcomes shown in Figures 9-3 and 9-4.

Very different patterns, are they not? In Outcome 1, there is a very simple main effect of stress shown in the left diagram and of success on the right. Outcome 2, however, shows that the effect of high stress— as opposed to low—is to *increase* persistence in subjects who have recently experienced success and to *decrease* it in those recently sub-

FIGURE 9-3. *Graphical representation of Outcome 1, showing two main effects and no interaction.*

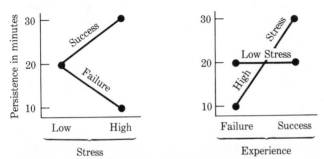

FIGURE 9-4. *Graphical representation of Outcome 2, showing interaction and one main effect.*

jected to failure. It shows also that the effect of an experience of success —as opposed to failure—is to increase persistence dramatically in the high-stress group but to make no change at all in the low-stress group.*

We might be tempted to speculate about possible explanations for those results. But in the first place this is a treatise on statistics, not psychology, and in the second place, these data were not derived from any real experiment in the first place. The point of it all in this context is that interactions do occur, and that when they do, they can be detected by analysis of variance.

Summary

Chapter Eight showed how a difference between two groups of subjects can be evaluated in terms of the probability of its occurrence by chance (sampling error). Chapter Nine has extended that technique to experiments in which *several* groups vary on a given factor (independent variable)—and even to some in which several factors must be evaluated simultaneously.

The logic of the *F ratio* is revealed in the course of its application to a *one-way analysis of variance* in a six-category experimental design. An example is given also of a more complex design to which analysis of variance might be applied—a two-way analysis in a 2 × 2 *factorial design*. Because *F* functions initially as a kind of "survey" test, the follow-up problem—the identification of the precise source of the variance when *F* has proven significant—is discussed briefly.

Finally, reference is made to Appendix I, which attempts to convey an appreciation of the breadth of application of significance tests and, simultaneously, to provide you with a ready reference list, by naming a large number of such tests and identifying their functions.

But the section described in the preceding paragraph is a departure from the major thrust of this book. Our objective throughout has been the understanding of basic concepts and their interrelations—the "big ideas," so to speak, in statistical thinking. In the last two chapters, for example, the "big idea" was the logic of testing for the significance of differences, especially in experimental settings; in Chapter Five, it concerned the description of relationships among variables in settings other than experimental. The first four chapters established a necessary foundation for everything that was to follow; but in Chapters Two, Three, and Four, it became apparent that "big ideas"—for example, the idea of "distribution," the importance of specifying the kind of average that is being used, the concept of "variability"—are involved even in the description of a single sample. Chapter Six stressed the importance, when interpreting any measurement of a given subject's behavior, of comparing it to similar measurements that have been made of some identifiable reference group. The central concern of Chapter Seven was the relation between a sample and its population.

All of those are important ideas. They won't enable you to *do* social science research, but they will give you access to research done by others. That is clearly a high enough payoff to have justified all the effort you have expended in mastering them; but there is an additional reward that, though incidental to the stated objectives of the book, could well prove most important of all in the end. While learning to think better *statistically*, you may have learned to *think* better!

EPILOGUE

You may not have occasion to "think statistically" every day, every week, or even every month; but whenever the occasion does arise, you will be ready for it. After an especially long hiatus, you may not be able to deal immediately with every concept that you have mastered while studying the book; but you will find that a quick reference to its discussion of a particular target concept will revive your understanding not only of that concept, but of many related ones as well.

General agreement is so far lacking on a single set of symbols to refer to the various concepts that have been developed here. For that reason, I have provided a conversion table (Appendix II) that will allow you to interpret any of the commonly used statistical symbols that you may encounter in your reading.

Good luck! And remember what Sir Francis Galton used to say: "Whenever you can, count."

TESTS
OF
SIGNIFICANCE

Virtually any kind of a difference can be tested for statistical significance. The only requirement is that the data be expressed numerically. Much of the research—especially the experimental research—that concerns social scientists and professionals is evaluated in those terms.

This appendix lists a large number of significance tests. Not all of them are labeled as tests of "differences"; they are such, nonetheless, because they are all testing the hypothesis of no (null) deviation from some specified point of reference, and the question in each case is whether the obtained deviation differs from zero.

Do not allow yourself to be frightened by the following formulas. Look at them first with the idea of discovering how broad is the application of significance testing; that is, indeed, the major reason for including them here. Later, you may wish to use this list as a handy reference, as you encounter various significance tests in your reading. Some of the tests are described partly or wholly in words, others entirely by their formulas. If the formulas—or the verbal descriptions, for that matter—don't make sense to you, don't worry about it. They will become more meaningful within the context of a research report. On the other hand, if you enjoy a challenge, you might want to try reading and conceptualizing the formulas (that is, comprehending the relationships expressed in them). I have provided some assistance in the form of labels for unfamiliar terms.

In all formulas, the effects of sample sizes on variance estimates are represented by degrees of freedom (see page 113) rather than N.

THE t TESTS

1. Correlation (r)

$$t = r\sqrt{\frac{N-2}{1-r^2}}$$

2. Partial correlation (of two variables, with the influence of one or more others partialed out)

$$t = r_{12\cdot3}\sqrt{\frac{N-3}{1-r_{12\cdot3}^2}}$$

($r_{12\cdot3}$ = Correlation of variables 1 and 2, with variable 3 partialed out)

3. Rank-order correlation (ρ, or rho)

$$t = \rho\sqrt{\frac{N-2}{1-\rho^2}}$$

4. Difference between correlated correlations (r_{12} vs. r_{13})

$$t = \frac{(r_{12}-r_{13})\sqrt{(N-3)(1+r_{23})}}{\sqrt{2(1-r_{12}^2-r_{13}^2-r_{23}^2+2r_{12}r_{13}r_{23})}}$$

5. Difference between regression coefficients (β, or *beta*)

$$t = \frac{\beta_1-\beta_2}{S_{d_\beta}}$$

(Standard deviation of differences between *betas*)

6. Difference between obtained and hypothesized ($\bar{X}_h$) means

$$t = \frac{\bar{X}-\bar{X}_h}{\text{SE}_{\bar{X}}}$$

7. Difference between independent means

$$t = \frac{\bar{X}_1-\bar{X}_2}{\sqrt{\dfrac{S_1^2}{N_1}+\dfrac{S_2^2}{N_2}}}$$

8. Difference between correlated means

$$t = \frac{\bar{X}_1-\bar{X}_2}{\text{SE}_{\bar{X}_1-\bar{X}_2}}$$

9. Difference between independent variances

$$t = \sqrt{\frac{S_1^2}{S_2^2}}$$

10. Difference between correlated variances

$$t = \frac{(S_1^2-S_2^2)\sqrt{N-2}}{\sqrt{4S_1^2S_2^2(1-r_{12}^2)}}$$

THE *F* TESTS

1. Difference between variances of two distributions

$$F = \frac{S_1^2}{S_2^2}$$

2. Analysis of Variance

(*a*) Treatments (entire matrix)

$$F = \frac{S_T^2}{S_e^2}$$

(Treatment variance, entire matrix)
(Error variance, entire matrix)

(*b*) Main effects (by factors)

$$F = \frac{S_A^2}{S_e^2}$$

(Factor A variance between A groups averaged over B levels)
(Error variance within A groups)

(*c*) Interaction (entire matrix)

$$F = \frac{S_I^2}{S_e^2}$$

(Interaction variance, various combinations)
(Error variance, entire matrix)

(*d*) Trends of means

1) Linearity

$$F = \frac{S_L^2}{S_e^2}$$

(Linear component)
(Error variance, entire matrix)

2) Curvature

$$F = \frac{S_Q^2}{S_e^2}$$

(Quadratic component)
(Error variance, entire matrix)

3. Eta (η) correlation coefficient for curved regression line

$$F = \frac{\eta^2/G - 1}{(1-\eta^2)/(N-G)}$$

(G = Number of groups)

4. Multiple correlation (R) of one variable with a weighted composite of two or more others

$$F = \frac{R^2/m}{(1-R^2)/(N-m-1)}$$

(m = Number of constants in the regression)
$R^2_{1\cdot23}$ = (Multiple correlation of variable 1 with 2 and 3)

5. Beta (β) weight in a multiple regression equation

$$F = \frac{R^2_{1\cdot23} - R^2_{13}}{(1-R^2_{1\cdot23})/(N-2-1)}$$

6. Coefficient of concordance (W) (Correlation coefficient indicating agreement among judges)

$$F = \frac{W(m-1)}{1-W}$$

(m = Number of judges)

NON-PARAMETRIC OR DISTRIBUTION-FREE TESTS†

1. Chi Square (χ^2).

2. Friedman analysis of variance by ranks (two-way).

3. Coefficient of Concordance (agreement of judgments).

4. Kruskal-Wallis analysis of variance (one-way). Called "H" test.

5. Median test. χ^2 used after division of groups at median.

6. Sign test. Positive vs. negative changes when experimental subjects are compared to matched control subjects. (Null hypothesis = same number of positive as negative changes.)

7. Wilcoxon test. Positive vs. negative changes, as above.

8. Mann-Whitney U test. Alternative to the t test when populations lack normality of distribution and/or homogeneity of variance.

9. Kilmogorov-Smirnov test. Goodness of fit of a theoretical to an observed frequency distribution. Alternative to chi square when N is small.

---

†The tests in this section are used (1) when the data come from discrete (as opposed to continuous) categories or (2) when quick estimates are needed. When these tests are used with continuous data there is a loss of information similar to that which occurs when the Spearman rho is used in place of the Pearson r as a coefficient of correlation. (See pages 37–38.) The tests are called "distribution-free" because they do not involve any assumptions concerning the distribution of the reference population; whereas, with the single exception of Chi Square, all of the significance tests discussed in this book were developed originally on just such assumptions. Since those assumptions now appear to be far less important than originally thought, the recent trend toward increasing use of distribution-free tests is currently being reversed.

MULTIMEAN COMPARISON TESTS

These tests are used after a significant main-effect F in analysis of variance.‡

1. *Duncan's multiple range test.* Means are arranged in a hierarchy, and ordered comparisons made (largest minus smallest mean, largest minus second smallest, and so on). Significance is based on the number of ordered steps between means. It is possible to test any and all means, taken two at a time. Each comparison must be planned in advance.

2. *Newman-Keuls method.* Ranked means or sums of experimental interest are tested for significance, based on the number of ordered steps involved in the comparison. Planned in advance.

3. *Studentized Range statistic.* Test for significance of largest minus smallest means only. Planned in advance.

4. *Dunnett test.* Comparison of each mean with a control or standard mean.

5. *Scheffé test* for any and all mean or sum comparisons. Does *not* need to be planned in advance.

6. *Orthogonal test* for comparison between independent sets of means. Planned in advance.

7. *Tukey tests.* Tukey has devised several techniques for locating the source of significance in an array that has produced a significant F. For details, see Tukey, J. W., "Comparing individual means in the analysis of variance," *Biometrics*, 1949, vol. 5, pp. 99–114.

---

‡Each of these tests is designed to ascertain which differences among groups are significant.

APPENDIX II

LIST
OF
SYMBOLS

The table on the facing page shows the symbols used in this book as well as alternative notations you may encounter in other statistics textbooks. Though the symbols used in this book may be the most common, there is no standard usage either accepted by authors in the field or approved by the American Statistical Association. [Personal correspondence with John W. Lehman, Executive Director, American Statistical Association, 26 June 1972.]

SAMPLE STATISTIC		POPULATION PARAMETER		NAME OR DESCRPTION
USED IN THIS BOOK	SOME ALTERNATIVE NOTATIONS	USED IN THIS BOOK		
X				Raw score
x				Deviation score
Σ				Summation
N	n			Number of observations
$\bar{X}$	M, x	μ		Mean
S	s, SD or SD	σ	†	Standard deviation
S^2	s^2, V, σ^2	σ^2	†	Variance
$SE_{\bar{X}}$	$\sigma_{\bar{X}}, \sigma_M, s_{\bar{X}}$, SE or SE	$\sigma_{\bar{X}}$	†	Standard error of the mean
$SE_{\bar{X}_1 - \bar{X}_2}$	$SE_X, \sigma_{(\bar{X}_1 - \bar{X}_2)}, \sigma_{d_M}$	$\sigma_{\bar{X}_1 - \bar{X}_2}$	†	Standard error of a difference
ρ			‡	Rank-difference correlation
r			§	Product-moment correlation
z t				Ratio of a quantity to its standard deviation or standard error
F				Ratio of one variance to another
χ^2				Chi square
p	P			Probability of wrongfully rejecting the null hypothesis. Level of significance.

†Only verbal reference is made to this concept in the text, but there is a fairly well-established convention that this parameter symbol should be used as a companion to the *statistic* symbol in the first column.

‡This system of notation calls for a transformation to Greek; since the statistic ρ is already Greek, that can't be done here.

§Here the system calls for a ρ; but that would be confusing, because ρ already has another meaning.

NOTES

The following supplemental notes are indicated in the text by asterisks.

Chapter Three

Page 14
A figure like this—essentially a vertical bar graph—is called a *histogram*. A histogram is one way of representing a frequency distribution.

Page 16
The distribution of means would be normal even if the samples were not.

Page 18
If you should ever have to do the computations yourself, you would find that the median is often *within* a class interval rather than precisely between two intervals as in the above illustrations. In that situation, it is necessary to interpolate; any methods text will quickly tell you how.

Page 19

A complication arises when the distribution includes the scores of two "types" of individuals. That can result in two or more modes (in which the distributions are called *bi*modal, *tri*modal, etc.). Multimodality should never be interpreted unless the N is rather large, however; for in a small distribution, chance deviations can easily produce a "humpy histogram." ("Histogram" is defined above in the first note on this chapter.)

Chapter Four

Page 25

Many statisticians use the quantity $N - 1$ in place of N when computing sample variance. (Sample variance is the square of the standard deviation.) $N - 1$ gives the number of *degrees of freedom* in the calculation of the statistic. (In other applications, there might be $N - 2$, $N - 3$ or $N - n$ degrees of freedom.) The calculation of the standard deviation requires the previous calculation of a mean, and one degree of freedom is "used up" thereby.

Imagine a very simple situation in which the individual scores that make up a distribution are 3, 4, 5, 6, and 7. If you are asked to tell what the first score is without having seen it, the best you can do is a wild guess, because the first score could be *any* number. If you are told the first one ("3") and asked to give the second, you probably won't guess some number in the millions, but logically there is no reason why it should *not* be in the millions or even higher; it, too, can be any number. The same is true of the third and fourth scores; each of them has complete "freedom" to vary. But if you know those first four scores (3, 4, 5, 6) *and you know the mean* of the distribution (5), then the last score can only be 7. If instead of the mean and the 3, 4, 5, and 6, you were given the mean and, say, 3, 5, 6, and 7, the missing score could only be 4. In any case, if the mean is known, the missing score is *determined* by your knowledge of the other four, and the *degrees of freedom* of all the scores taken together has been reduced by 1; hence the term "$N - 1$."

Our calculation of the mean was done directly from independent samples of the population (see pages 13–17). The number of degrees of freedom was therefore equal to N. But when we work with deviation scores, as we do in the average deviation and the standard deviation, degrees of freedom is less than N because the deviations are from a mean, and the calculation of a mean uses up one degree of freedom. (See above.) Since all our statistics are based upon the assumption of independent ("free") observations, any statistic that does not make a

correction for loss of independence, or "freedom," is a biased estimate
of its population parameter. (Again, see pages 13–17.)

In this book, that bias will be overlooked, and N will be used everywhere as degrees of freedom. There are four reasons for so doing. The
first is that the main text carefully develops the concepts "mean of the
raw scores"

$$\left(\frac{\Sigma X}{N}\right)$$

"average deviation"

$$\left(\frac{\Sigma |x|}{N}\right)$$

"standard deviation"

$$\left(\sqrt{\frac{\Sigma x^2}{N}}\right)$$

and finally (on page 94) "variance"

$$\left(\frac{\Sigma x^2}{N}\right)$$

as a sequence of constructs characterized by a common structure; to
have changed N to $N - 1$ in the second formula would have broken
the sequence. The second reason for omitting that refinement is that
it does not contribute to an understanding of any of the abovemen-
tioned statistics as measures of variability. A third reason for leaving
it out is that even writers of more advanced texts often use

$$\sqrt{\frac{\Sigma x^2}{N}}$$

as their measure of variability in the sample, with or without apologies
to their colleagues. And finally, if degrees of freedom $= N - 1$, as is
the case of every statistic discussed in the main body of this text, the
difference between "N" and "degrees of freedom" is negligible, providing only that the sample is not extremely small. (With a sample of
only two cases, for example, N would be *twice* as big as $N - 1$.)

Page 27
The semi-interquartile range is a much more precise statistic than the
total range, but no adjustment is made *there* for the $\frac{1}{2}$ unit cut off at
either end of the Q_1 from Q_3 interval in the process of subtracting Q_1
from Q_3. In the case of the total range, my own view is that the only
time that such an adjustment is really necessary is when the question

to be answered is specifically, "What is the smallest interval within which all cases will fit?" If you were to refer to "elementary school grades one through six," for example, it would be misleading to speak of a range of five grades (largest = 6, minus smallest = 1).

Chapter Five

Page 31

Actually, a Pearson r (pages 34–42) coefficient on these data would be slightly less than 1.00. The relationship is perfect, but r is accurate only in rectilinear (straight-line) regressions, and this one is really curved. There is another coefficient that can be used in such cases; look for "eta (η) coefficient" or "correlation ratio" in any standard text on statistical methods.

Page 35

Those assumptions are: (1) that each of the two distributions is unimodal and symmetrical, and (2) that the line best representing the relation between them is straight rather than curved. (See the diagrams in the Scatterplot section of this chapter, pages 37–42, for examples of such "rectilinear" relationships. Also see the preceding note concerning curved ones.)

Page 43

A sample must be taken at random or by some other approved method; but even when the "simple" random method (see Footnote 1, page 4, for definition) is not used, randomness characterizes an important part of the total selection procedure.

Page 57

Originally, a T score was any standard score other than a z score, but usage has gradually changed its meaning to "a standard score in a distribution that has a mean of 50 and a standard deviation of 10."

Page 63, first note
Usually median.

Page 63, second note
And his IQ is 100. (Divide mental age by chronological age,

$$\frac{\text{MA}}{\text{CA}} = \frac{6}{6} = 1.00$$

Then drop the decimal point).

Page 63, third note
And his IQ is 133 (8/6 = 1.33).

Page 63, fourth note

See "Deviation IQs" in Figure 6-3. The IQ defined in the second note (above) for page 63 was for many years *the* IQ. Now, however, it is known as the "ratio IQ" to distinguish it from the newer "deviation IQ." (See "Deviation IQs" scale in Figure 6-3.) A deviation IQ is a standard score. In a distribution of general mental ability test scores, the obtained mean is converted to 100, because the mean ratio IQ is 100. Then the standard deviation of the raw scores is changed to match that of ratio IQs on the same test. The result is a set of standard scores that are almost the same as would have been obtained by the old method. The new method is easier to use, and it has other advantages the explanation of which would require a more extensive digression into psychometrics than would be appropriate here.

Chapter Eight

Page 82

This is the formula that is used when means are uncorrelated. If students were *matched* on some variable—for example, intelligence—that is related to their level of performance under both control and experimental conditions, a correlational factor would have to be introduced. But that refinement is beyond the scope of this book, and the basic idea of

$$\mathrm{SE}_{\bar{X}_e - \bar{X}_c}$$

is better conveyed by Formula 8-1 as it stands.

Page 83

To find how many standard errors an obtained difference is from zero, it was necessary to find the size of the standard error. In this instance, there were two such computations:

First, there was the standard error of the difference for Figures 8-3 and 8-4, assuming that the standard errors of the means of the two distributions are as indicated in Figures 8-1 and 8-2:

$$\mathrm{SE}_{\bar{X}_c} = 2; \ \mathrm{SE}_{\bar{X}_e} = 2$$

$$\mathrm{SE}_{\bar{X}_e - \bar{X}_c} = \sqrt{\mathrm{SE}^2_{\bar{X}_c} + \mathrm{SE}^2_{\bar{X}_e}}$$

$$\mathrm{SE}_{\bar{X}_e - \bar{X}_c} = \sqrt{2^2 + 2^2} = \sqrt{4 + 4} = \sqrt{8} = 2.83$$

$$\mathrm{SE}_{\bar{X}_e - \bar{X}_c} = 2.83$$

Second, there was the standard error of the difference for Figures 8-7 and 8-8, assuming that the standard errors of the means of the two distributions are as indicated in Figures 8-5 and 8-6:

$$SE_{\bar{X}_c} = 5; \ SE_{\bar{X}_e} = 5$$

$$SE_{\bar{X}_e - \bar{X}_c} = \sqrt{SE_{\bar{X}_c}^2 + SE_{\bar{X}_e}^2}$$

$$SE_{\bar{X}_e - \bar{X}_c} = \sqrt{5^2 + 5^2} = \sqrt{25 + 25} = \sqrt{50} = 7.07$$

$$SE_{\bar{X}_e - \bar{X}_c} = 7.07$$

Page 85
The "significance level" states the probability that we are making an error when we reject the null hypothesis. That kind of error is known as "Type I." By stating in advance of the experiment that we will accept only a very low probability (that is, a low probability of rejecting a null hypothesis that is in fact true), we can reduce our Type I errors to a level approaching zero. Unfortunately, however, the *lower* we set *that* level, the *higher* is the probability of *accepting* a null hypothesis that is in fact *false*. The latter kind of error is called "Type II." The following table summarizes those relationships. The ability of a significance test to make the decision represented by the lower left cell of the table—that is, to reject a false null hypothesis—is called its "power."

DECISION

		Reject	Accept
	True	Type I error	Correct
NULL HYPOTHESIS	False	Correct	Type II error

Page 86
Many authorities will not accept a one-tail test under any circumstances. Their argument is that we cannot have a good reason for expecting one group to be superior to the other—or rather, that if we did we wouldn't have to make the statistical test. To put it another way, the reason that we conduct the test at all is precisely that we *don't* know what the outcome will be. There is controversy in every field.

Chapter Nine

Page 89

In our examples, the "no difference," or "no deviation," hypothesis specifies no deviation from a difference of zero between (or among) treatment groups. Although that is far the most frequent usage, it is possible to *begin* with samples from different populations and try by means of the experimental treatment(s) to increase or decrease the already existing differences among them. In such a case, the null hypothesis would be that "the original difference will remain unchanged" in spite of the treatment(s).

Page 90

Technically, of course, it is the null hypothesis that is being tested rather than the obtained difference. But when such a test reveals that the null hypothesis is probably not true, most researchers allow themselves to speak of the *difference* as being signicant.

Page 93

The computation may be of secondary concern to some readers. Here it is:

$$\chi^2 = \Sigma \frac{(f_o - f_e)^2}{f_e}$$

$$\chi^2 = \frac{(50 - 60)^2}{60} + \frac{(50 - 40)^2}{40} + \frac{(70 - 60)^2}{60} + \frac{(30 - 40)^2}{40}$$

$$\chi^2 = \frac{-10^2}{60} + \frac{10^2}{40} + \frac{10^2}{60} + \frac{-10^2}{40}$$

$$\chi^2 = \frac{100}{60} + \frac{100}{40} + \frac{100}{40} + \frac{100}{40}$$

$$\chi^2 = 1.67 + 2.5 + 1.67 + 2.5$$

$$\chi^2 = 8.34$$

Table III of Fisher, R. A., *Statistical Methods for Research Workers*, Edinburgh and London, Oliver and Boyd, 1932, gives 6.635 as the χ^2 required for significance at the .01 level; so our 8.34 is significant at an even higher level—that is, at *less* than .01.

Page 94

An alternative to the S^2 notation is V. But there is a more important alternative to the formula itself: if you have read the note (above) for page 25, you know that the quantity used as the denominator is often $N - 1$ rather than N. (See the last note under *Chapter Eight*, above.)

Page 99, first note

When there are only two groups to be compared, the analysis of variance is really a *t* test; or to put it another way, *t* is a special case of analysis of variance that can be used only when there are just two categories of the variable being examined. In such a case, $t = \sqrt{F}$; with large samples, of course, the same is true of *z*—that is, $z = \sqrt{F}$;

Page 99, second note
The *F* ratio for interaction is

$$F = \frac{S_I^2}{S_E^2}$$

where S_I^2 = interaction variance and S_E^2 = error (within-group) variance, both derived from the entire matrix.

Page 101
In order to achieve maximal simplicity—and hence clarity—in this example, I have assumed the linearity of each of the functions depicted in Figure 9-3 and 9-4. A more refined experimental design might have defined those functions more precisely by including *several degrees* (instead of just two) of stress and *several degrees* of success. If that were done in the case of stress, for example, it might turn out that *medium* amounts of stress produce more persistence than either extreme amount. An investigation of the effects of a single variable through many changes in its value is sometimes referred to as a *parametric* study.

SUGGESTED READINGS

There are many valuable texts that discuss the application of statistical principles to problems in business, economics, education, psychology, sociology, and other related fields. Here is a very small sample of what is available:

Champion, D. J. *Basic statistics for social research*. Scranton, Pa.: Chandler, 1970.

Dornbusch, S. M., and Schmid, C. F. *A primer of social statistics*. New York: McGraw-Hill, 1955.

Downie, N. M., and Heath, R. W. *Basic statistical methods*. 3rd ed. New York: Harper and Row, 1970.

Freeman, L. C. *Elementary applied statistics: For students in behavioral science*. New York: John Wiley and Sons, Inc., 1965.

Games, P., and Klare, G. *Elementary statistics: Data analysis for the behavioral sciences*. New York: McGraw-Hill, 1967.

Glass, G., and Stanley, J. *Statistical methods in education and psychology*. Englewood Cliffs, N.J.: Prentice-Hall, 1970.

Guilford, J. P. *Fundamental statistics in psychology and education*. 4th ed. New York: McGraw-Hill, 1965.

Klugh, H. *Statistics: The essentials for research*. New York: Wiley, 1970.

Mason, R. D. *Statistical techniques in business and economics*. Revised ed. Homewood, Ill.: Richard D. Irwin, 1970.

Shao, S. P. *Statistics for business and economics*. 2nd ed. Columbus, Ohio: Chas. E. Merrill, 1972.

Willemsen, E. *Understanding statistical reasoning: A guide to the critical reading of the research literature in the social sciences*. San Francisco: W. H. Freeman and Co., in press.

INDEX